Rebirth of the Small Family Farm

A HANDBOOK FOR STARTING A SUCCESSFUL ORGANIC FARM BASED ON THE COMMUNITY SUPPORTED AGRICULTURE CONCEPT

THIRD EDITION

by
Bob and Bonnie Gregson

ACRES USA
THE VOICE OF ECO-AGRICULTURE

Foreword

Two people can make a reasonable, community-oriented, non-exploitive, earth-friendly, and aesthetically pleasing living on a very small farm. And what value can be assigned to holding a job you love, in a wonderful environment, with someone you love to be around, while providing a valuable service to your community? Add in unlimited continuing education in all subjects from botany to plumbing, ever-changing challenges, much time to widely think and contemplate while performing quiet tasks, lots of exercise and fresh air, the best food available anywhere on the planet, a 30-second commute, no "dress for success" costs, and the responsible flexibility that comes with being in charge of your own activity.

This kind of farming can even be considered a privilege.

— Bob and Bonnie Gregson

Contents

Introduction

I t is, poetically, a dark and stormy October night as this (true) tale begins to be set on paper. We have just shed our raingear and come inside after a session of slaughtering nearly 100 over-the-hill chickens. They are easy to snag from their henhouse perches after dark, and it was slightly comforting to do this periodic unpleasant-but-necessary chore in the rain.

How bizarre this thought process — let alone the actual deeds — would have seemed to either of us eight years ago! We were then in our mid-40s, had just remarried and bought a small operating flower and nut farm on an island near Seattle, and were determined to drop out of corporate careers for a better lifestyle.

So we did. And life did get better, much better.

But getting from there to here — with two very nice little businesses built up on the premises — was extremely challenging. What follows is about the process and what worked for us. We believe it can be easily adapted and improved nearly anywhere even by novices as we were.

In a Nutshell, So to Speak

We named our potential piece of paradise Island Meadow Farm . . . because it had some meadowy areas and is on an 8 mile by 15 mile island, a 20-minute ferry ride to the windward side of Seattle. Nearly 100 years old, it consisted of 13 acres, a turn-of-the-century "fixer-upper" farmhouse, numerous outbuildings in various stages of disrepair, a dozen plastic hoop houses, and a century's worth of "this might come in handy someday" junk piled everywhere. In addition to its southern exposure and wealth of horticultural diversity, its greatest asset was its potential. We were extremely optimistic; we just didn't know where to begin.

Nowhere in our research could we find broad and reasonably universal guidelines for most start-up aspects of a very small farm featuring just two people — we already knew we did not want to depend on outside labor sources. In fact, the conventional wisdom was (and is) that, aside from a few unique situations, a very small farm simply can't make enough money to

We were extremely optimistic . . . we just didn't know where to begin.

4

support a family in a reasonable way, so there is no point in trying. We were so intent on making our farm our main livelihood that we forged ahead anyway.

Contemporary authors Eliot Coleman, John Jeavons, Andy Lee, Shepherd Ogden, Joel Salatin and Gene Logsdon did offer substantial hope for what we wanted to do, though their experiences tended to involve hired and/or apprentice labor. Their books are a "must read" for potential new small farmers, and provided very valuable guidance as we struggled along.

From those sources and many others we synthesized ideas; after eight years we now recognize our combination of farming activities to be a financial and personal success. Unlike some other small farm success stories, this farm operation is just built around wholesome food — not fads. And it is based on hard work and direct marketing; there is almost no possibility that someone can franchise the concept and live off the hard work of others!

You Can Do It!

What's really most significant is the apparent universality of this model. It is easily replicable throughout most of this country and many others, making good use of small acreages near urban areas.

If we had known what we know now from day one on the farm, we very likely would have been financially successful by the end of the third year instead of the eighth. And we are confident this can be accomplished by "informed novices," with relatively small capital expenditure.

So here's a bit of the background story and a blueprint of what works here.

We offer this book from the perspective of two people working together to make a reasonable, community-oriented, non-exploitive, earth-friendly, and aesthetically pleasing living on a very small farm, in the fervent hope and expectation that many thousands of these 2- to 10-acre farms will pop up and be successful all over the country . . . with many fewer trials and errors than we experienced!

1
Our First Years on the Farm

Sometimes it takes failures to push you in the right direction.

Our combined farming experience amounted to Bob's working in the eastern Washington wheat harvest during a high school summer, and Bonnie's weeks spent on her grandparents' Iowa farm at age ten. In other words, no experience. We both have almost always raised some vegetables and fruit trees in our back yards. (In fact, Bob has the distinction of growing what has to have been the closest personal garden, ever, to the Pentagon, while living in Army housing only a few hundred yards from that establishment in 1967-68.)

Each of us spent nearly 20 years in various nonagricultural business management positions before taking up farming. That background helped in many ways, giving us the skills to think in economic terms, develop a business plan, and recognize the importance of proper marketing. Our broad-brush approach was quite good. Here, verbatim, is our first statement of "Aims" and "Business Concepts," which have since proven to be mostly valid.

These broad concepts really did and do make sense. The problem came in the details.

Aims

1. To have a comfortable living
2. Not to be constantly tied down
3. To create a place of beauty — orderly
4. To produce healthy things
5. To have a lifestyle that is "sharable" with others
6. To have a lifestyle which minimizes stress
7. To have flexible schedules
8. To have fun

Business Concepts

1. Retail on site
2. Operate professionally
3. Capitalize on existing assets
4. Produce and sell year-round
5. Operate without borrowed funds
6. Ease into more food production (Note: it was flowers and nuts at that time)
7. Develop specialties (freesias, herbs, orchids?)
8. Minimize labor
9. Optimize land use

These broad concepts really did and do make sense. The problem came in the details.

The Original Business Plan

The business plan we developed was simply a naive sales forecast, using the rosiest data to ensure that it would show we could make a living on the farm. For example, we extrapolated average western Washington yields from (presumably "normal") apple trees and bee hives, and where local data were not available, adjusted national averages for things like nut tree production, taking no account of the condition of the trees. We projected flower sales based

6

on sketchy information provided by the prior owner. That person was seemingly making a living for himself and a somewhat communal group of helpers; we figured we could do what they were doing and enhance other aspects.

So we struggled along with our business plan, trying to make the farm work as it had during the previous decade, raising flowers and some nuts, and continuing some of the wreaths and dried flower arrangements successfully done during the prior regime. It was quickly obvious that the numerous ancient unkempt fruit trees and disreputable beehives were not commercially productive. It was also apparent that the local market had become flooded with dried flower crafts.

Inexperience Can be Humorous in Retrospect

Once, during this time, we sold a flat of plums to a similarly struggling friend who was making and selling pies for a living. The plums were a red, clingstone type from an old tree. And they tasted good fresh. Bonnie made a cobbler with some of them the next day: it was horribly sour and awful, despite normal sugar input. We immediately contacted the pie maker to warn her. Too late! She had already made and sold the plum pies . . . "Pies", her customers let her know, "from Hell!"

But the real "killer of the business plan" — of course unknown to us when we bought the farm — was that a plant disease called botrytis had recently infected all the flower fields and greenhouses, dooming all future bulb crops, with no realistic organic or chemical cure. Since those crops had been a primary money-maker for the prior owner, we had counted heavily on them in the plan.

Similarly, the seven acres of filbert and walnut trees were way beyond their economic life and their output barely paid the property taxes in a good year. So the business plan was discovered to be best suited for composting by the end of the first year.

A New Look

We knew we loved to grow edibles best. And we also knew we would grow organically for philosophical and practical reasons. So we were pleased when a successful specialty (organic) salad grower decided to lease some land from us and move her operation here. This turned out to be one of many wonderfully serendipitous events leading to where we are. She shared information and we could watch and learn from her practices.

In the meantime, our farm business was going nowhere. The upholstery business Bonnie had bought and quickly learned prior to our marriage plugged along on the premises. It couldn't support us completely; besides, she wanted to be growing things, not working in a shop.

Bob was fixing up the house and other structures, sorting through mountains of junk left all over the property, growing some flowers and produce to sell on weekends at a Seattle farmers' market, and operating a small gift shop we opened in one end of the shop building. Nothing much was happening as envisioned in the business plan!

The financial situation grew so bad that we sold other assets and bought a photography franchise. That turned out to be a wasted year, but not a disaster, because we were able to sell the franchise back. As the photography experience became obvious for what it was, we concluded, most unhappily, that we had to sell our beautifully restored farmhouse with all of the buildings and some of the farm property to carry us over into the next (unknown) phase. That painful decision was another serendipitous choice of those early years. We profitably sold the house and three acres, paid off the land, and installed a manufactured home, pole barn, and small homemade greenhouse in the middle of the remaining ten acres.

Another Beginning

That relocation hurt our pride but put us in excellent financial condition — and moved us from the edge of the property to the center. Living right next to the most productive fields

Grape vine growing on plastic mesh.

This small farm model built around subscription farming is very nearly universal, widely replicable, and minimally subject to whims of the market.

The garden gate.

has made a huge difference in many ways. It is aesthetically pleasing to look out on our beautiful growing area and it saves much time going back and forth for tools, supplies, and so on. "Out of sight, out of mind," is a reality with an impact... as our farthest fields can attest.

Bill Mollison's "Permaculture" concept came into play here: one should, according to his ideas, have an overall farm design that places those crops requiring the most attention closest to the farmhouse. So we put most of the vegetables and all the berries around the new house/shop/greenhouse area. A new fruit orchard and space for undemanding vegetables like winter squash and potatoes went into the south slope field. The old nut trees stayed where they were, though some near the house were recently removed for vegetable growing space.

As this was happening, our specialty salad friend was facing her own challenges.

She decided she needed a break from the business after that second year here (her seventh or eighth year in the salad business). We agreed to take her equipment and remaining supplies and customer list in exchange for that year's land lease plus some cash. Presto, we were suddenly specialty salad wholesalers!

And Yet Another Beginning

Her business was focused on upscale restaurants; at that time the fancy salads with edible flowers and various unusual greens sold for sixteen dollars per pound. (A few years before, in the early '80s, it was often twenty or more dollars per pound.)

Our first full year in that business was a good one, shipping to her former customers all over the Northwest. It was so good we were able to have several backyard organic growers grow greens to help us meet the demand. We were all tooled up for the next year, and had others ready to join in again.

The End of That Beginning

But early that next year large mechanized California growers entered the local wholesale market and the wholesale price for a similar — but not nearly as varied or fresh or pretty product — dropped to four dollars per pound. Just as we were preparing to deliver the first seasonal batch of salad to our largest customer, they abruptly told us they wouldn't be using our product any longer!

That's how institutional buyers operate unless under contract. Either a chef suddenly changes — as they routinely do about once a year at most restaurants — or they find a better deal somewhere else and feel no loyalty to the producer. It was a terribly hard lesson. That year was another financial struggle since we had lots of product but only one-third of the past year's market; and it was a disappointment to the several others who were planning to sell greens to us.

We limped along for a few weeks, wondering what to do, but even at the time recognized that each "cosmic kick in the pants" had led us in new and more productive directions. What would be next?

The answer wasn't long in coming, and was, in fact, already on hand.

A New Way to Connect With Customers

For over a year a friend, a former organic farmer, had been sending us information about a new way to market direct to the ultimate consumer, something called "Community Supported Agriculture." It was being tried on the East Coast and even at several farms in Washington. The concept generally involved retail customers coming directly to a farm to pick up a prescribed amount of produce.

We highly value our privacy; it did not look inviting to consider further marketing schemes wherein the customers would come right down the driveway to the farm. The gift shop operation at the original farmhouse was not a particularly happy experience, requiring too much time with customers who were just enjoying the outing and the interesting handicrafts, and wanting to look at the animals or have a picnic or . . . whatever.

It took the shock of losing the bulk of our new business to begin thinking that maybe our privacy would have to suffer! We pondered what this new way of doing things might mean. Then decided to give it a try.

During those decision-making weeks we joined a new farmers' market in Seattle, wondering if it would work better for us than the other had. It did. It also bought time and created new exposure to help start our version of Community Supported Agriculture, a subscription farm operation.

The Subscription Approach

So after those first four years of fits and starts in several directions we hesitatingly began what is known as "Subscription Farming" as the core of our business. It means providing fresh produce all summer long on a prepaid basis to a group of customers. And it means much more than that on a deeper level: it frequently seems to connect people to their food and to the land in a mysterious way that helps satisfy some important inner instincts. Perhaps it is just a "grounding" in more ways than one, in a world that doesn't feature much of that anymore.

Whatever the reasons, it has been a great success for three full seasons and promises to be so indefinitely.

Subscription farming works so well as the core of our business, and for many other resurgent small farms in this country, that we feel obligated to share the concept and its complementary parts. We want to do this especially for those, who, like us, somehow know that they want to live a life more connected to the essentials of earth, sky and the recesses of one's own soul — but don't have a clear idea of how to begin.

Here is the best step-by-step way we see to make that beginning. And it must be methodically done. Starting out "cold" in subscription farming is a ticket to failure.

Begin Almost Anywhere

This small farm model built around subscription farming is very nearly universal, widely replicable, timelessly attractive, and minimally subject to whims of the market.

It's time to throw in a special comment here. What we've described of our first four years may, by this point, sound gut-wrenching and like an awful experience. It had a number of those moments as described, plus the tragic death of one of Bonnie's adult sons, ongoing concerns about Bob's disabled son, petty struggles with a former spouse, and occasional damage to our bodies. But it was rarely boring, always challenging, frequently fun, and the most deeply satisfying career either of us had ever known or could imagine. And it still is!

It's Worked for Us Novices

We believe *you* can farm successfully almost from the beginning if you (1) have a strong work ethic and desire, and, (2) rigorously base your activity on certain key factors.

This book will cover those factors. It will not describe how to grow crops. Many good books and periodicals comprehensively do that. This book will describe how to grow an economically successful old-fashioned mixed crop farm on as little as two acres.

A Two-Acre Farm?

"A farm? Two acres can be called a farm?" some may ask. If a person spends full time tending livestock and working the soil by hand to produce numerous food crops for sale, he/she is as much a farmer as ever has been. In fact, that person is much more like the time-honored farmer than most "agribusiness operators" of today. So we say, without apology, that even two acres can assuredly be a farm in the traditional and best sense of the word.

We believe you can farm successfully almost from the beginning if you have a strong work ethic and desire, and rigorously base your activity on certain key factors.

2
What Our Farm Does: An Overview

Two acres provide our living and are the focus of this book.

The term "Subscription Farming" may be unfamiliar to many. It is one of several variations on a relatively new concept called Community Supported Agriculture (CSA). That concept calls for direct linkage between the person who grows the food and the person who eats the food.

Some CSA farms are pure shareholder operations, where everything grown is divided equally among the shareholders, and the shareholders often provide some of the labor. Some CSA farms are the result of a group of people (parishioners of a church, for example) hiring a farmer, providing land, and dividing the output among themselves.

The CSA option most prevalent on the West Coast is the subscription type, like ours, where the farmer prearranges season-long weekly produce deliveries to a group of prepaid customers.

Advantage to the Consumer

The end result is similar in all cases: the consumer gets the freshest possible produce, knows something of the ethic with which it is grown, and gets a small sense of what farming is all about — the good and the bad — during the 20 or more weeks of seasonal communication and connection with the farmer.

Advantage to the Farmer

A prepaid, predetermined "general" market is the main advantage to the farmer. The term "general" here delineates a real advantage: the farmer gets to choose what goes to the customers each week. This gives the farm operation huge flexibility; it encourages the farmer to try new varieties and new growing techniques without fear of catastrophic monocrop failure, since numerous crops and varieties keep the subscribers happy.

Some crops will do well each year and some will not, depending on weather and other factors; failures will accordingly be offset by successes without recourse to crop insurance or government programs. The weekly offering to the subscribers must be varied and interesting, so the consumer has seasonal surprises to look forward to each week.

Growing a wide array of crops adds much complexity to the farm operation. On the other hand, it fits within an attractive format of personal relationship and quality that cannot be duplicated by large farms (though some are trying). It would be prohibitively expensive for large farms to hire the dedicated, multifaceted management and workforce required to maintain quality at levels easily obtainable by the two of us on our small acreage. The bottom line is that our type of farmer cannot be whipsawed between major suppliers and major wholesale markets.

Competition

What about competition? It is quite different for the farmer in this setting. In the wholesale arena, price is usually the essential element, and small farmers rarely can price-compete with large mechanized operations. If the small farmer sells retail, he/she either has a farmstand and competes somewhat with local grocery stores and other farmstands, or he/she sells at a farmers' market. Both of those are very competitive activities.

The consumer gets the freshest possible produce, knows something of the ethic with which it is grown, and gets a small sense of what farming is all about.

Subscriptions, on the other hand, once established, are rather like a marriage: the farmer has a whole season's guaranteed sales to the subscriber. If he/she does a good job of providing, the subscriber will be back next year and will tell friends about what a great thing the arrangement is, so a large part of each year's sales are guaranteed in advance.

The parties thus enter into a business relationship built on trust, human relationship, and mutual interest. It establishes a mutual loyalty and connection similar to that which our American cultural myth says existed in "the good old days" before corporate America took over Main Street.

Subscription farms compete with each other to a very limited extent, mostly when signing up the first batch of customers. The ongoing primary advantage over other marketing options, as noted above, comes into play again: having a full year of a relationship gives the customer a very good idea of the quality, quantity, and overall value of the relationship, and it gives the farmer a sales year without competition.

Since the farmer's personality determines the nature of his/her complete product and how it is presented, each farm program is surprisingly different from others in the area. Customers either relate well to the whole package, or don't. Since each of us tends to attract, and be attracted to, "our type of person," the field is thus wide open for little clusters of people doing primary business with "their farmer" over a period of years. (We have overheard customers refer to us as "their farmer"!)

Note: *In a greater metropolitan area of 2-3 million people, if one out of every 20 households participated in a subscription program, it would take about 1,400 farms like ours to meet the demand! What would be the economic and aesthetic value to the whole community if 1,400 farm families and their related infrastructure were busily prosperous on the periphery of that community?*

Getting Started

How did we start? We lined up 20 subscribers for the 1993 season (and will describe the process later). That number is not magic, but seemed a good starting point based on our experience growing salad greens. It also fit with our desire to make the operation work without dependence on hired help or others.

No Hired Help

That latter point needs explanation. We choose to do all the work ourselves, which limits our scope, but is an educated choice for profitability and simplifying our lives. Our salad predecessor, and other small farm operations we know about, grossed a substantial amount of money but netted very little because of high labor costs; this is very common with all farms. It is usually difficult to find, and almost impossible to adequately pay, experienced farm workers, given the generally low prices paid for farm produce relative to the cost of living.

That factor, plus the extra office and paperwork time involved when one has employees, and the time required for proper supervision, led us to conclude that it is not only simpler and more socially benign, but also more profitable to do the work ourselves. Besides, we had spent over 35 years between us in managerial roles, and were quite ready to get away from the stress and strain of dealing with purely "people" kinds of issues.

Since the farmer's personality determines the nature of his/her complete product and how it is presented, each farm program is surprisingly different.

11

How It Went

The first year with the 20 brave subscribers went fairly well. Only a few dropped out at the end of the year, for various reasons, so we replaced them and added more from an accumulating waiting list to reach 30 subscribers the next year as we gained confidence; then decided on 35 as the maximum after that — but took a few more, and it worked out fine.

In addition, we sell from a small self-serve farmstand along our driveway, and some to a florist, a local restaurant, and grocery store. These fit together nicely. Our first objective is to provide adequate quantities of produce to our subscribers. If there is a surplus, it goes to the farmstand, where it is available to any customer, including subscribers who want extras. If it looks like there will be more than the farmstand can quickly sell, the surplus goes to the grocery store. Some things like Delicata winter squash, Sugar Pie pumpkins, sweet onions, hazelnuts and tomatoes we purposely grow in much larger quantities because they are in demand (at a good price) at the grocery store, and are relatively easy to grow. The same general idea is applied to flower sales.

Farm Layout

Geographically, our Island Meadow Farm is a gently south-sloping ten acres of diversely poor Maritime Northwest soil. Glacial remnants provide for interesting digging! The 800-foot

utilities trench from the county road down our driveway traversed everything from iron-red sand to boulders to blue clay to gravel. We count on about five inches of topsoil throughout, but are often disappointed.

Most of the land is orchards of very old, decaying filbert/hazelnut (essentially interchangeable names) trees; about two acres are cleared for cropland, green-houses, house, shop, compost pile, grape vineyard, fruit orchard, ponds, chicken houses and outbuildings.

Those two acres provide our living and are the focus of this book.

A simple, attractive sign tells the community who you are and where you are.

3
Key Factors to Success

Small farms are a labor of love.

As one considers what to do to make a living on such a two-acre parcel, there are major factors to consider, a screen through which one's dreams must pass.

The ten key factors we have identified, and discussion of each, follow:

1. Like-minded partners — mutual support is crucial.
2. Organic growing — ethics and economics happily coincide.
3. Keeping it small — a few acres are enough.
4. A passion for growing — small farms are a labor of love.
5. Physical ability — are you a willing worker?
6. Ongoing education — knowledge is the key.
7. Direct marketing — cut out the middlemen.
8. Location close to consumers — nearby population is an asset.
9. Value-added product(s) — two plus two can equal five.
10. Off-season work options — have a backup skill.

1. Like-minded Partners

We believe it takes two like-minded, very supportive people to make most start-up enterprises work. And a small farm, like any small business, is indeed intense during the early years! It will require all the focus the partners can give.

All successful subscription farmers should be out standing in their field.

That is not to say that it is impossible to do this with a family or one partner's career and its related needs — but the partners must mutually understand, appreciate and fully support the motivation involved in operating the farm, and be prepared to sacrifice most other outside activities during the early years.

It is obvious to us that our coherent interests are in large measure responsible for our success. To be blunt with a personal subject, it is also obvious that this would not have been the case with our prior spouses; neither of our prior marriages would have withstood this activity level because the spouses did not have the same passion for walking/talking/working/reading about and thinking about growing things, and, quite understandably, could not have supported or substantially contributed to the overall effort. And neither of us could likewise contribute to their passionate endeavors.

Remember, we said that this is a great life and very fulfilling, but probably only if this is where your passion lies. It is crucial to identify personal life passions in advance. Dragging a partner into an intense field of endeavor into which he/she does not fit may be fatal to the relationship.

13

2. Organic Growing

Small farms should grow crops organically. The more we read and discover, the more we see the long-term disaster that chemical farming is, both for the consumer and especially the farmer, and we also see how feasible it is to grow mixed crops better organically. (We have closely followed the debate on this topic for many years, and have experienced it ourselves, so do not make the previous statement lightly.)

Over 70,000 chemical poisons have reportedly been used in recent years in various phases of agriculture; most are still in use. No one knows how any of the 70,000 or their by-products interact with each other in the various soil-groundwater-air combinations, let alone how they impact the 100,000 or more living entities found in each teaspoon of topsoil. Or what impacts chemical combinations have on humans. There is no way a rational, dispassionate observer can assume toxic agricultural chemistry is benign to humans; after all, it is designed to kill living entities.

Farm workers and farm families have many horror stories about their personal experiences with those toxic chemicals. Many such persons are killed and damaged around the world every year.

Those poisons plus synthetic chemical fertilizers are also a huge expense to farmers. Fortunately, they can be replaced by much cheaper homemade items, especially on the small farm where chickens and other farm animals can play a substantial role.

Save Money and Build Healthy Soil

So why would anyone use expensive products that are undeniably dangerous to the applicator, to the eater of the produce, and to all nature, if they are not necessary?

Because well established multi-billion-dollar financial and bureaucratic institutions are dependent on their sales and use, that's why. They have created a dependency and a mindset that seemed to make sense for awhile after World War II, but is now known to be dangerous and unnecessary in almost all cases. And they focus on symptoms instead of the real problems.

These powerful institutions are stridently promoting their chemicals and "silver bullet" procedures at all levels, and have the financial power to strongly influence the political and educational system.

If the buggy whip industry had enjoyed equivalent financial power early in this century, during the transition from horses to automobiles, it is not unreasonable to believe that every new car would still have a federally-mandated buggy whip as standard equipment!

Many farms, both large and small, have decades of successful organic production of a wide array of crops. Good crop rotations, feeding the soil with natural amendments and compost, and good choice of varieties replace toxic chemistry and serve everyone better in the long run.

Productivity of Organic Farms

It is sometimes said, even recently, that chemical farming is dramatically more productive than organic. That is utterly false. For a two-year sample study of twenty eight farms see the article by Klepper et al in the February 1977 *American Journal of Agricultural Economics,* showing approximately equal performance in Midwestern commodity crops.

Another myth alleges that organic produce is usually small, ugly and blemished. It can be — but won't if well grown and properly cared for. Another myth is that pests run rampant through organic fields. The truth is that healthy soil grows healthy, pest-resistant plants, and encourages a natural balance of so-called pest and predator.

Some organically grown produce actually tastes better, too, especially carrots, beets, potatoes and walnuts. Good organic farmers grow beautiful, high-yielding, tasty produce.

14

If you need further reasons for using the best of the old-fashioned ways, consider this: organic produce is also a market niche commanding a higher price and increased consumer interest/loyalty.

Chickens and Other Farm Animals Belong in the Scheme

As a corollary, we think organic farming almost has to include raising truly free-range chickens, birds that have complete access to varied outdoor areas.

True free-ranging chickens convert various leftovers, crop residue, general forage, and purchased feed into terrific input for the compost pile; they devour most weeds, weed seeds and bugs they find; they aerate and till the soil in a healthy way (as long as they aren't left in the same place too long); and they provide eggs, or meat, that, again, is notably better than what you can find at the grocery store.

Our eggs are a real draw to our farmstand because they are so fresh, with an orange, upright yolk that tastes appreciably better than the "factory" eggs from stressed, caged chickens. It is usually impossible to find grocery store eggs from free-running chickens, despite the clever — but meaningless — marketing words like "range," "ranch," "naturally nested," and "free-range" that adorn many commercial egg cartons. Farm stand eggs will be, at most, several days old; those in the grocery store may be several *months* old.

Ethical Considerations

People are also beginning to have ethical concerns about how production animals are treated; many like to know they are patronizing a farm that lets chickens and other farm animals freely run in a congenial setting.

Final Note About Organics

Replacing synthesized agricultural inputs with meticulously defined organic inputs is a huge step in the right direction. But it is only one step. We firmly believe that it is equally important to use diversity, intercropping, companion planting, "friend strips" (habitat left for insects), cover cropping, crop rotations, wind breaks, runoff collection, and integrated livestock techniques to achieve the robust balance of life in and above the soil that enhances all agricultural activity. People and communities also count in the equation: ethical, caring attitudes should be a hallmark of everything we do when we call ourselves "organic growers."

3. Keeping It Small

Two acres, or just a few more, are desirable. More than that requires substantially more capital and more labor than two people can provide in an intensive-growing situation.

Most of our income — not counting egg production — comes from 3,400 square feet of raised beds plus about 27,000 square feet of general growing area. That is less than three-quarters of an acre. And we are not nearly as "intensive" in that area as we could be with ever-better planning and implementation.

We've already noted our conclusion that two supportive partner beginners can make more money on small acreage than can be made hiring help and using more land. Each situation might unfold differently, but, for starters, we think it is best to limit production activities to one-half acre.

"Acreage Creep" is One of the Downfalls of the American Family Farm

It is most seductive to think that cropping a few more acres will justify purchase of a machine that will do things faster. That's probably true, the part about the machine doing it faster. But machines are expensive, need shelter, insurance, maintenance, and fuel and usually only relate to a few crops. Extra acres and their necessary machines mean more debt. They also automatically mean the farmer will not be able to pay as much detailed attention to each

We believe it takes two like-minded, very supportive people to make most startup enterprises work.

growing area, so quality will diminish as time demands increase. The larger size may not be at all justified in the overall sense.

Everything we see tells us that economically successful farming is easiest on the very small or very large farm. We know about other newcomers like us who have expanded with more acres, machines, and employees, but seem to have a great many more headaches — and working hours — without any more net income, and perhaps less, than provided by our few producing acres.

Income Expectations From Those Few Acres

Let's be a little more specific about income. In our King County, which includes Seattle, the median household income (half are above this, and half below) is about $54,000. Use that as a benchmark for adjusting the following table to economics of your area. And note that you should *net about 75%* of the gross sales.

Don't expect the farm to totally support you during the first several years. Our experience, and that of other growers, suggest the following (1995 dollars):

Year	Gross Sales	Sources of Income
1	$10,000	Farmers' Market / Friends
2	15,000	Farmers' Market / Friends Grocery/Restaurant/Florist
3	20,000 +	Subscribers Grocery/Restaurant/Florist Farmstand
4	30,000 ++	Subscribers Grocery/Restaurant/Florist Farmstand Plus Other Niches You Discover

"Acreage creep" is one of the downfalls of the American family farm.

Our largest expense, by far, is about $3,500 annually for . . . chicken feed (300 chickens)! . . . the next biggest is about $400 for seeds. Both can be greatly reduced by better farming techniques such as seed-saving and better use of the orchard floor (growing food crops there for the chickens).

For the two-acre, general-purpose farm model operated by two people, there appears to be a practical limit in the $40-50,000 net income range. Interestingly, that latter figure will probably put them in the top 5% of all U.S. farms, and they will make a better living than most farms much bigger and busier.

Our thinking and practices evolve as we gain experience: we currently see some new ways of doings things that will move us toward that income range within several years. And there are plenty of examples of people who have found very specialized niches like ginseng, baby vegetables, specialty flowers, or houseplants, that can generate a great deal more money and profit on several acres. "Windows of opportunity" open with experience.

Income Comparisons

Comparing $30-40,000 to what we jointly made at corporate jobs years ago, and what salary ranges are in the corporate world today, this farm income is indeed modest for two workers! And there are no medical/dental or retirement or vacation benefits, and that amount won't service much of a mortgage.

But — we're talking about a seven- or eight-month program per year in this climate zone. Taken in that light, it's not so bad. And, what dollar value can be assigned to holding a job you love, in a wonderful environment, with someone you love to be around, while providing a valuable service to your community? Add in unlimited continuing education in all subjects from botany to plumbing, ever-changing challenges, much time to widely think and contemplate while performing quiet tasks, lots of exercise and fresh air, the best food available anywhere on the planet, a 30-second commute, no "dress for success" costs, and the responsible flexibility that comes with being in charge of your own activity. This kind of farming can even be considered a privilege.

Biologically Sound, Productive Farming Begets True Wealth

Speaking of wealth, is rich topsoil a prerequisite for a farm site? No. Good biological farming soon begets top-notch soil.

Gathering organic filberts by hand.

Your two acres does not have to start with rich topsoil. Ours certainly did not. It would be helpful, but not essential, except for a fast start-up. Soil can and must be built over time; a lot can be accomplished in three or four years.

The "lay of the land," length of growing season, and immediate microclimate are crucial. It is advisable to discuss these issues early on with knowledgeable locals. Those issues will establish what you can best grow, and where, and when.

Farmers and observant gardeners know that there are substantial climatic ranges in a locality, and even within a smallish garden. Close observation will tell what each little zone is like under various weather circumstances, and, thus what type of plant will grow best in each "microclimate."

4. A Passion for Growing

Potential farmers need a burning desire to grow things, a labor from the heart. Life on a small farm can be intense, wearying, and sometimes boring or frustrating. One has to have the internal fire that stays lit despite setbacks. It is essential to honestly ask oneself hard questions about personal strengths/weaknesses. Am I willing to carefully plan? Do I organize well? Do I work well under time pressure? Do I work well on my own? Do I focus and complete projects?

Neither of us had seriously explored the possibility of making a living in agriculture. Sure, we knew great happiness and satisfaction came from working in the garden — our own gardens — but no connection with a livelihood in our own gardens was obvious. Mid-life crises apparently jarred each of our synapses enough to complete the "why not try this?" circuit.

It seems that the internal "growing" fire is often lit by grandparents long ago, particularly by nature-conscious grandmothers, and may lie smoldering for many years before it becomes obvious and acted upon. That was true for both of us.

The test for this, or any other potential life passion, lies in the answer to the question, "What activity would I most often enjoy doing if it was totally up to me to decide?" Most of us do not consciously know the answer to that important question and do not consider relating it to our professional lives. That is a tragedy on a grand scale!

Don't expect the farm to totally support you during the first several years.

What would be the effect on the collective mental and spiritual health of our culture if even one of every ten adults was passionately, lovingly pursuing a career he/she felt was his/her special calling?

5. Physical Ability

Ability — and the willingness — to work long and hard physically is also crucial. These two items are a bit different from each other and should be scrutinized accordingly.

Bob forking hay from the garden cart.

Ability to Work

The work required on a very small farm is within the capability of most people, male or female, large or small.

The two of us led relatively sedentary prior lives up to our mid-40s so it took some time, and painful experiences, to adapt to a steady diet of physical exercise. Our biggest physical problem in the early farming years was that we did not quit doing something when it started to hurt. Long periods of repetitive motion were the most dangerous.

Bob spent nearly three incapacitated weeks after three days of steady dirt shoveling. His left shoulder sort of burned after the first day, then hurt continuously as he dug the next two days, then hurt enough by the end of the third day to stop almost all motion. Yes, of course that sounds dumb, and it was. He was still thinking about how he could do things 20 years before, with no after effects.

Bob learned two main things from this experience: how to temporarily function with only one arm; and, not to trust the "work through the pain" ideas so dear to some of his more macho Army experiences years ago. It's not the same body!

Willingness to Work

Willingness of the mind may be a different matter. We, for example, typically and (usually) cheerfully each work 80 hour weeks throughout the growing season. Even at that there is always a backlog of unfinished business. Those with small children or other time-consuming interests might find that routine unacceptable.

And sometimes, as noted earlier, the work is hard, boring, tedious, frustrating, painful, or all of the above. Sometimes it is glorious, beautiful, spiritual, joyous. Regardless of what may prevail at any given time, the totality is immensely rewarding to those tuned to the right frequency, and well worth the labor of love. But that labor is relatively unceasing; one cannot expect to be financially successful "dabbling" in small-scale agriculture.

6. Ongoing Education

The 80-hour weeks mentioned above can be minimized through improved techniques and timing. Some of that comes from personal experience; better yet, and usually much cheaper and less frustrating, is to capitalize on others' experience.

Read. Listen. Read. Ask questions. Read some more. You get the idea.

Read Books

Most of our successful activities and techniques come from books we've read. After all, several thousand years of cumulative agricultural experience rests in society's collective memory, including the written word. Just a few decades ago, one could find all kinds of books full of wise farming observations, or look just about anywhere in this country and find an old, experienced farmer, and tap into a fairly consistent ancient knowledge. Reading is important; getting wise personal advice is also important, but hard to come by.

Good Advice is Hard to Find

First, it's hard to find any farmers nowadays. There aren't many left.

Second, the experienced farmer of the late 20th century is a product of post-World War II chemical farming practices. He/she knows all about operating large equipment on a large expanse of land, using procedures developed by and, in effect, dictated by the chemical companies, the federal government, and his/her banker. That may sound overly simplistic. It isn't really, when you consider the long-term interaction of corporate grant money with the agricultural universities, corporate money in politics, and the revolving job door among public agricultural officials, agribusiness and agri-education.

This is the new collective knowledge: how to deal with government programs, how to use high-tech equipment, how to negotiate with marketing middlemen, and how to work with a distant banker. Even the fertility and pest control program, and crop choices, are often in the hands of those same outsiders or hired consultants (who work for the very firms that will sell and/or automatically apply the remedies they prescribe).

Our modern farmer is minimally aware of the complexities of the living soil, spends little time with hands or feet on the soil, and observes few of the many nuances of plants and soil so prominent to his ancestors. Farm family "farming" skills have been replaced by reliance on technocrats and bureaucrats.

Some farmers understand this; some resent it; to most it is just a fact of life.

All of this is just a long way to say that there is not much next-door help available to any of us as we work to resuscitate the best of the time-honored farming skills. Fortunately, many old books offer valuable information and there are some terrific new books. Those at the top of our list include:

The New Organic Grower, by Eliot Coleman.

Backyard Market Gardening, by Andy Lee.

These provide excellent details about small mixed crop operations. They both seem to be widely available at book stores.

The Organic Method Primer, by Bargyla and Gylver Rateaver.

This is the ultimate book about how to grow any food crop organically. It is expensive but is worth every penny.

Salad Bar Beef, Pastured Poultry Profits, You Can Farm and *Family Friendly Farming,* by Joel Salatin.

The several basic texts about "Permaculture," by Bill Mollison.

These cover the concepts of farm layout in the broadest sense, insuring that everything fits together in a multipurpose, mutually supportive sort of way.

Read Periodicals

We subscribe to, and thoroughly read, about a dozen somewhat specialized periodicals, plus others of general agricultural interest. The most valuable ones are listed in Appendix C.

7. Direct Marketing

A small farm must primarily sell directly to the consumer.

To make a living on a small farm, we must cut out the middlemen who today make virtually all the profit between farm harvest and ultimate sale to the consumer. Subscription farming and good use of a farmstand allow the farmer to sell produce for substantially more than the normal wholesale price.

This kind of farming can even be considered a privilege.

Example: If we grew summer squash on a large scale, early in the season it would sell in bulk for about 35 cents per pound to a local wholesale warehouse. Or we could sell lesser quantities at about 55 cents per pound direct to a few organic-friendly local grocery stores who would price it to the consumer at about $1.00 per pound. Or we could sell even smaller quantities directly to consumers at our farmstand for about 75 cents per pound. The latter is clearly "price optimal" for us and the consumer, cutting the consumer's produce cost by 25% while raising the farmer's gross by 36 to 114%. Those are big numbers in aggregate.

Advantages of Direct Sales

The farmer makes more money, while the consumer spends less money. That was clear in the example.

Organic produce display at a Seattle Saturday market.

But there are other factors not so readily obvious: food travels the 100 feet from the field to our farmstand in several minutes, with much less handling than during its four or more days from a faraway farm to a wholesaler to a grocer and then to the consumer. That means our model provides much fresher (translated as more nutritious) food with less wear and tear on the interstate highway system, less fuels burned to transport, handle and preserve the produce, less cultural stress on our social system since two-person farms are not dependent on migrant workers, encourages use of more interesting varieties (not just bred for easy transportability) within plant families to protect genetic diversity, reduces the need for energy-intensive food storage facilities with attendant chemical preservatives, and keeps more dollars in the immediate local economy.

Those add up to a huge plus for the taxpayer/consumer.

Disadvantages

There are really two main drawbacks to direct sales: first, it takes time and effort to establish/maintain sales facilities; and, second, the quantities that can be sold by the farmer are much less than from wholesale opportunities.

Time and effort are the principal inputs on a small farm, so we have to be careful that we don't overextend our human hours. The farmstand at our Island Meadow Farm is self-service and only one hundred feet from our house, yet it still requires 20-30 minutes per day for restocking, sign-making, checking status of change on hand, and so on. And building a pleasant, attractive facility takes time and resources. Our little stand cost about $500, and the 800-foot driveway plus parking area requires about $200 worth of new gravel each year, but these are well worth the expense.

Quantity of Product Available for Direct Sale is a Real Issue

For example, it is not reasonable to even consider selling truckloads of wheat from a farmstand. Nor will one normally sell thousands of pounds of any one thing. So the small farmer realizes that along with the higher selling prices possible through direct sales comes a smaller sales volume potential. The clever farmer plans accordingly, ideally growing enough of each crop to just saturate the highest price market available (the farmstand customer group), with maybe a little left over to sell to the next-highest price buyer (usually a local grocery store or restaurant.)

8. Locate Close to Consumers

A small farm should be within a reasonable distance of potential customers. Towns offer wide varieties of food consumers in a rather dense cluster. Since everyone's time is at a

The wise old farmer, embodying the agricultural wisdom of the ages, is almost an extinct species.

premium, logic dictates farming within a short distance of a city or town where one can quickly deliver to the consumers or they can easily come to pick up produce and see the farm. If one out of every 20 families would participate in your farm program, it only requires a population base of 2,000 families to make that program successful. About 100 cars come down our driveway to the produce stand each week, probably representing 100 families.

Your customers will come to consider you "their farmer." Most will love the opportunity to see — and have their children see — where their food comes from.

9. Value-Added Products

A small farm must feature at least one frequently-used value-added product. "Value-added" means adding labor to a basic item in such a way that the end result is worth more than the beginning raw material. Apples, for example, may be worth 50 cents per pound raw, but much more as dried fruit or after conversion to fancy vinegar and packaged/bottled in an attractive way.

Many don't understand this concept. For example, our four-state agriculture newspaper ran a January 1996 article titled, "California Economist Debunks Value-Added." This "marketing expert" went on to say that the popular new salad mixes in grocery stores weren't really successful because they weren't increasing sales of lettuce. Let's hope she was misunderstood: The point of value-added is to add two dollars worth of product to two dollars worth of labor and then sell the new product for five dollars instead of the traditional four dollars.

No farmer should be in the business of "selling more lettuce"; he/she is in the business of growing and selling things in the most creative way to maximize profitability. Selling a good salad mix is much more profitable than just selling more lettuce.

The keystone of small operations is that through personal attention one can grow beautiful produce, but this will, by definition, be in fairly small quantities. Therefore one must carefully choose those products that make the most economic sense.

Choosing Produce to Grow

We focus on higher-value items for the farmstand: tomatoes, winter squash, raspberries, broccoli, pole beans, and our value-added specialty, a 60- to 90-ingredient gourmet salad mix. That list might be quite different were we in another area or climate zone.

Sure, we grow enough of everything else under the sun to keep it interesting for our subscription customers, but there is no way we would consider growing corn for sale at our farmstand where we would have to price it at eight or more ears for a dollar to be locally competitive. Nor would we grow radishes or something that takes substantial labor and space, yet sells for pennies per bunch.

The salad, on the other hand, is a way to make good money from humble components. It is highly popular because it's fresh, includes interesting different tastes, and is ready to eat out of the bag. We add value by picking it leaf by leaf, tearing leaves to bite sizes as needed, mixing 60-90 ingredients together, double or triple washing the mix, and ensuring a variety of colors, tastes and textures in each bag. At ten dollars per pound, it sells on a continuous basis to a large group of year-round loyal customers who say they are now "addicted." It accounts for nearly 30% of our yearly gross revenue.

Beware of Fads

Is our salad mix a fad? It probably was to some extent when the edible flowers were considered the essence of its marketability. Our customers now tell us — and we listen closely for this type of feedback — that they like our salad because it is obviously fresh/healthy, tastes so much better than the iceberg and — in adventurous families — romaine lettuce salads they grew up on, changes seasonally, and is ready to eat direct from the bag.

Gregsons' Theorem:

Growing and selling only basic produce will assure the small farmer a poverty level income.

21

Flavor, interesting variability, healthiness, and "ready to eat" are factors that will only become more important as our go-go world keeps going faster.

Cosmos.

There are other farm products that have come and gone in our regional popularity polls. Some kinds of "cutesy" things are especially vulnerable, items that are not eaten or frequently used, like dried flower crafted arrangements, some herbal products, and edible fresh flowers.

Rule of thumb: if you start reading about a trendy new non-staple product in your regional lifestyle magazines, the product market base is already saturated — you're too late. Find or create some other niche.

10. Off-Season Work

Lastly, it is a time-honored tradition for small farmers to have an off-season way to make money. This is important as a financial supplement during the start-up years and a refreshing change of pace, if desired, later on. Bonnie bought a small upholstery business and learned the trade from the prior owner before we bought the farm. It's a congenial, warm and dry business we can operate at one end of the shop building during the cold winter months. By mid-October we are ready to shift gears from the growing season, and yet stay productive; by early February we are eager to start growing again. It's a nice combination. Here again, one must be creative to identify skills that can be tapped for an on-farm job. Skills like carpentry, jewelry-making, various crafts and computer based talents are good candidates. The Amish people represent good models; read about what they do.

Summary: *Those are the ten most identifiable keys to our small farm success. You may be able to finesse one or more of the points. But beware; glossing over any with unrealistic bravado may shower the dream with burning arrows. Our original farm business plan certainly suffered from some arrow holes, as described earlier, but these ten major underlying conditions were either directly met or readily adaptable.*

4
Start-Up Requirements

About $11,000 will buy everything necessary to start the farm business.

If you have comfortably passed through the ten-point screen you are over the major hurdle. So what is needed to get a profitable small farm going? That's easy: farmers, land, some equipment and supplies.

Farmers

We've already talked about some of the characteristics of people who can make this work. There are few actual requirements. New farmers have to be hard-working, observant, focused, idea seekers, and learners. At least one of the group on your farm should have good "people skills," since much about a successful small farm operation involves friendly, positive customer relations.

We often say that we were a bit too old (45 and 42) when we started farming. But age is circumstantial. A couple could probably take over a well-established operation in mid-life or beyond if they were fit, there wasn't much development work to do, and they were otherwise prepared. Scott and Helen Nearing are fine examples of those vigorously active in later years; their books are exceedingly inspiring. They allegedly started the current back-to-the-land movement while well past mid-life, over 40 years ago.

At the other end of the age spectrum (20-30) danger may lurk in taking on something of this intense nature prior to having a wide array of life experiences — not being ready to leave the hurly-burly and really settle in. What seems like a refuge to many of us can seem a prison to others!

Land

There are so many variables about selecting land that we'll not add any more than noted in the earlier discussion about farm size.

Equipment

We started our dream fairly well capitalized. After a major house renovation . . . and hasty purchase of the wrong tractor . . . and wrong truck . . . and wrong irrigation system . . . we were much wiser and much less capitalized.

Farm equipment is one of those areas where many of us, particularly males, have built-in assumptions. We see ourselves on a large tractor doing "farmwork." The tractor probably is a serious looking older Ford or John Deere that we have miraculously purchased in mint condition, and intuitively know how to operate and maintain. Some may fantasize that they bought the tractor for a song, and rebuilt it in their spare time, using the new knowledge conferred on males when they achieve a rural route mail address.

Beware of That Tractor Syndrome

Tractors, of themselves, are nothing but a (very attractive) power source: the attachments are what do the work. Those attachments come in all sizes, shapes and functions. The farmer must be very knowledgeable to pick the right attachments for his/her operation. Those choices in turn more or less dictate the choice of tractor.

Farm equipment is one of those areas where many of us, particularly males, have built-in assumptions.

But you probably don't even need a tractor in the beginning.

Frankly, even many experienced farmers fall into the trap of buying unnecessary or too much equipment. One can usually rent machines or hire someone with specialized equipment when needed, saving a great deal of money in the long run.

It's really pretty simple. Don't buy anything until you know exactly why you need it, what it will do for you, and why something — or someone — else much less costly can't do the job.

Farm Trucks

Let's also counter tradition right away by stating that a truck or van is great, but you can get by with a sturdy car, especially a mini-van, in the early years. Our 1987 mini-van has handled everything from fifteen tall buckets of flowers to six adult sheep.

We recommend the flowers versus the sheep. The sheep had gone for a little outing after breaking through a fence, and were ultimately captured too far away to herd back home. A very bad afternoon. It almost raises a sweat five years later. "Lawnmower sheep" cost us far more than they were worth in the long run.

Can There be Farm Life Without a Pickup?

Despite the universal belief that a pickup is required, we have had several kinds and ultimately decided there is more value to this farm from a van. A van best protects produce, animal feed, straw, and so on, from rain, wind, sun and dust. Those items are the things most often carried, and the latter are the elements most dangerous to them. What a van cannot easily haul — the bulk items — we have delivered or make on site.

Novelty sunflowers are beautiful standing in the field and can be sold as cut flowers or dried for natural bird feeders.

One can put a canopy on a pickup to provide that same protection, it is true, but then you just have a van with less headroom and overall useful space. Also: good, tight-fitting canopies do not easily go on and off a truckbed.

Other Primary Necessities in the Maritime Northwest

There is a core of equipment requirements on a small produce farm, one which varies somewhat by region and by specialty. Make sure you have solid evidence and/or experience before you buy additional appealing items — and many temptations will surely present themselves.

Start-Up Costs

In our area, after you've arranged home, land and a vehicle, it currently (1995) requires about $11,000 to purchase all essential farming start-up items. To some that may sound like a lot; in farming terms it is unbelievably small change. Those items and rough cost include:

ITEM	ROUGH COST
Rototillers	$ 2,900
String Trimmer	350
Hand Tools	200+
Carts and Wheelbarrows	500
Temporary Shelters	700
Perimeter Fencing 1/2 acre	600
Cold Storage	900
Irrigation Equipment	700
Nursery Ground Cloth	400
Raised Growing Beds	200
Metal T Posts & Netting	400
Seeds	300
Washing/Mixing Tubs	100
Scales	100
Chicken Housing/Fencing	300
Miscellaneous Hoses, Materials, etc.	800
Soil Test & Amendments	300
Business License	100

The next chapter discusses each category in more detail.

Yearly Expenses

Following are some of the items you will consider in your yearly operating budget: chicken feed, straw, replacement chicks, water, seeds, soil amendments, insurance, fuel, electricity, vehicle costs, maintenance, depreciation, property taxes, dues, publications and professional education.

5

Overview of Equipment and Supplies

Know what you need, and then buy the best.

Rototillers

You need two rototillers: one good strong gear-driven 10-12 horsepower model and a very lightweight tiny one. We paid about $2,600 for the former (a BCS 10 horsepower) and $330 for the latter (Mantis).

These are the 4th and 5th tillers we've owned in recent years; they are the best we've found, but still aren't easy to use. Even the largest walk-behind tillers do not do a good job on sod or thick cover crops or straw mulch without making numerous passes. That's a problem for the soil and a hassle for the operator. It would be far better for the soil, and your body, to get it all done in one pass, without pulverizing the ground as happens in three or more passes.

Another tiller issue is that in some areas they may not be useful because of rocky ground. You may need a tractor to do any tilling. Find that out early on from others in the area.

Tillage is a controversial area in today's agriculture . . . the less the better, is the general rule, but it appears there is no reasonable way to get around doing some on our farm, getting rid of slug habitat and eggs, turning under cover crops, and controlling weeds. New European spading-type attachments for tractors and very large rototillers may be a good, but expensive, solution.

String Trimmer

A string trimmer is very useful for cutting tall cover crops, as well as general property trimming. The traditional hand-held kind, even the good ones, cause numbness in our trigger fingers/hands after 20 minutes or so of steady use. The wheeled version, like the "DR" machine (that's its trade name) advertised in various garden publications, works very nicely and saves your body pain. It costs about $350 for the three horsepower model.

Some enjoy using a scythe to do cutting and trimming work. We have found that virtually any repetitious motion for a period of time causes joint problems and should be avoided. Maybe that's just us.

Hand Tools

Hand tools are the heart of the program despite the previous comment. Get good ones . . . but not necessarily from fancy mail order houses. You need sturdy shovels and digging forks, a hay fork, standard hoes, a triangular hoe, a stirrup hoe, both sizes of Winged Weeder, heavy-duty rakes, a T-post driver that pounds in most, but not all, metal fence posts, a posthole digger, a maul, an axe, a pick-mattock, a five-foot steel railroad track bar, plus hammers and normal hand tools.

You will add to your collection as you gain experience. Wheeled weeders, seeding machines, sprayers, flame weeders and other devices should be tested later on as you determine a definite need.

Our experience with the small hand tools like trowels suggests that there are yet no truly ergonomically-correct ones on the market. They are all hard to use for extended periods of time, contributing to carpal tunnel problems we both have encountered. This is a problem waiting for a solution.

Hand tools are the heart of the program . . . get good ones.

26

Start-up hand tools will cost several hundred dollars. But you probably already own many of them. Do not plan to drive fence posts with a maul; dig in wooden posts and only use metal posts that a T-post driver can drive. Bob wasted much time and energy using a maul to drive metal fence posts in the early years here.

Carts and Wheelbarrows

Carts and wheelbarrows contribute mightily. Again, get good ones. You will need a large-wheeled heavy duty cart and a plastic or steel-bodied 5^3/$_4$-cubic-yard wheelbarrow. The plastic body will probably last longer than steel in wetter climates; the steel may hold up better if you must carry lots of big rocks. These will likely cost close to $500 total.

Temporary Shelter Structures

You will want two 12' x 40' plastic-covered temporary shelters as appropriate for the weather conditions in your area. One can be a potting shed, plant starting area, and protected storage area; the other should be for growing high value crops, allowing you to extend the temperate season in spring and fall.

There are many variations on how to build these. They are temporary structures, usually not requiring a building permit, and you can find plans in many garden books. Look around at what other growers use in the area since wind and snow conditions determine type of construction. Plastic covers also vary; review the options and look at local examples before choosing. An average type may cost $300-400 for materials. You can certainly build it yourself.

Find out what your local animal pest problems are and what fences will work to solve them.

Most experienced small-acreage commercial growers eventually build more substantial replacements that fit specific needs and cost a lot more.

Perimeter Fencing

Animal fencing can save much grief and money. Our particular nemesis, aside from our former demented sheep and current chickens who occasionally get into the wrong areas and wreak havoc, is a large local deer population. There are many theories how to deter deer and other wild critter depredations. Don't waste time and effort testing them; build a good fence around your growing area(s) from the very outset.

Find out what your local animal pest problems are and what fences will work to solve them.

We knew we had to have a barrier over eight feet tall to discourage deer, so built a rather inexpensive fence featuring hog wire on the first three vertical feet, then plastic netting for three more feet, then old telephone lines festooned with surveyor's tape at the eight-and-a-half foot level.

Posts are about ten feet apart. Every third post is a 10-foot treated wood post sunk about eighteen inches into the ground. The two in-between posts are 7-foot heavy steel "T" posts pounded in about one foot, anchoring the hog wire and the lower portion of the netting; the higher part of the netting and the telephone wire are thus only supported every thirty feet.

Materials, not including the old telephone wires, cost about $1.10 per lineal foot for this fence.

When we bought the farm there were amazing piles of "stuff" scattered around the outbuildings, almost all overgrown with weeds and covered with more debris. An entire wood and glass greenhouse had been salvaged from somewhere, a boiler system and lots of steampipe from somewhere else, and massive amounts of other things were on hand. Most was so badly decomposed as to be unusable, even by a dedicated "collector of maybe useful things" such as Bob. We hauled many truckloads to the dump and to a metals recycler.

One huge pile — a mass of old overhead telephone wire and similar material — was the subject of much debate. What use could it ever have, and how could it possibly be untangled? Wouldn't it be better to take it to the recyclers who would pay for the copper or aluminum?

Well, we ultimately spent a solid two winter weeks unsnarling and rolling it into hundreds of rolls, and it has served us very well for fencing and raspberry wires — about one half mile used so far, and lots left in inventory!

Success

This tall but relatively flimsy setup has been successful in stopping deer, but does nothing to hinder pesky pheasants or squadrons of crows attacking corn seedlings . . . and the good old sheep did force under/through a gate once while en route to setting the kiwis back about two years. Even though sheep are aesthetically pleasing, they can be a real trial despite your best efforts to control/protect them. We no longer have sheep for these reasons.

Fence as Trellis

Fences also do multipurpose duty as trellising around the edges. You may want to grow beans, cucumbers, etc., on it, or use the fence as support for berry vines or sunflowers or . . . whatever else animals won't destroy from the outside.

Cold Storage

Cold storage is almost crucial. Some get by without it, and we did for several years before we did on-site sales, but it is so helpful that we strongly recommend it. Our 40-cubic-foot ex-Coca-Cola cooler, with glass sliding doors, cost about $900 used. It's well worth the money, and mandatory for a business that features salad greens and on-site sales.

We recently added a 7-foot by 15-foot homemade walk-in cooler that gives even more flexibility. It is not well designed insulation-wise, so don't ask us for ideas until we get it right!

Irrigation Equipment

Irrigation equipment takes away the threat of drought. T-tape, a trade name, is a drip tape system that has worked very well to avoid overhead watering and efficiently deliver water right where it's needed. West Coast sources and their fully descriptive catalogs are listed in Appendix A.

It's not uncommon to irrigate here from mid-April through early October, even in the Maritime Northwest popularly known for rainy weather. Correct watering is so important that we would never attempt to rely on local rainfall. It's an insurance policy; we would certainly install a drip tape system for any intensive growing operation, anywhere. The tapes are easy to lay and remove, are inexpensive, and may be reused for a number of years if carefully handled.

Our very flexible irrigation setup for one and a half acres cost $700 or so. Earlier efforts with the recycled rubber tire "leaky hoses" didn't work very well; they distributed water unevenly, broke if bent improperly (hard to avoid doing that), and clogged because of relatively hard water and algae buildup.

Ground Cover

Woven plastic ground cloth is a real labor saver, suppressing weeds between rows of beans, tomatoes, squash, and most anything else where rows are three to six feet apart.

It allows rain and air to pass to the soil but essentially stops weed germination. It is easy to roll out and recover after the season, thus lasting quite a few years. But it is expensive. Try several rolls. They are usually in three- and six-foot widths, three hundred feet long, and cost about five cents per square foot. Invest $300-400 in several widths after you have a growing plan and know how much pathway area could be protected this way.

28

Raised Growing Beds

Raised beds have many advantages. They are very useful for growing frequently-picked items, like salad greens. We have 800 lineal feet of four-foot-wide raised beds made with 2 x 6s, 8s or 10s, plus half of the fenceline around one 100 x 200 foot plot is thus "encased" to allow the fence to be better used as a trellis.

Working in Raised Beds

It's surprising how much wear and tear a small elevation saves on our backs during planting, picking and weeding operations. Other elevated advantages include faster warmup in the spring, clear delineation where the actual growing area begins — and thus where compost or nutrients should be placed — and a clear break point where grass from the paths may not enter.

Advantages of Using Hard Edges vs. Mounded Edges

These advantages are enhanced by having boards or something solid serve as bed edges; we had serious multi-season weed problems on edges of beds that were simply mounded up, and don't use that technique anymore.

Edge Materials

Treated lumber or plastic boards or rot-resistant wood for bed edges is prohibitively expensive. Also the treated lumber has some nasty side-effects . . . six chickens died here after ingesting some treated sawdust two years ago.

The cheapest solution we have found is to buy sling loads of mill-end random length utility-grade fir/hemlock boards, then paint them with a nontoxic mix of $1^1/_2$ cups linseed oil to 1 gallon turpentine, then nail on long strips of old greenhouse plastic to cover the inside and bottom edge of each board that would otherwise contact the earth.

Pathways Between Raised Beds

One more point about raised beds — how to treat the paths in between? If the beds are less than lawnmower width apart, we cover the path with woven plastic ground cloth. If easily lawnmower accessible, as they should be, the paths are left to natural grasses, which, when mowed frequently, produce terrific nitrogen-rich mulch and compost material.

We would certainly install a drip tape system for any intensive growing operation, anywhere.

Pea, Bean and Tomato Supports

Metal T posts (seven footers work best) and five-foot field fence serve as seasonal trellises for tomatoes and beans. Nothing else we've used has been strong enough. Posts may be at eight-foot intervals for peas and beans, four feet for tomatoes.

Seeds

Seeds cost about $300 per year. Patronize sources that are working to maintain heirloom, open-pollinated, and genetically diverse plant families. Sources for seeds are listed in Appendix A.

Wash Tubs

An old bathtub plus two washtubs, in which to wash and drain produce, cost under $100.

Scales

One or more scales are needed. If herbs are involved, you need one scale accurate from 1 to 32 ounces. Salad greens and produce usually are weighed in quantities up to five pounds; bulk items like fruit or potatoes go to thirty pounds or more. We eventually got a scale in each category.

Tilling under a cover crop.

Though it doesn't often come up, scales used in public selling situations like farmers' markets must be certified accurate by a local agency. Start with a used scale that is accurate and goes up to at least ten pounds. It will cost about $100.

Portable Fencing and Chicken Housing

Moveable fencing plus materials to safely house and feed twenty-five chickens, plus the initial chicks, will cost about $300. Several good reference books, describing how to design and build chicken facilities, are available at your local library.

Miscellaneous

Garden hoses, packaging materials, plug trays, potting soil, chick feed and miscellaneous other items you select will cost at least $800 for start-up. You probably own some of these already, and can get plenty of grocery bags and empty egg cartons from friends. We'll talk more about plant-starting materials in Chapter 9.

Let's put row markers in this general category. We staple 4" x 4" plastic cards onto 30" stakes then write on the plastic with indelible markers (which do fade badly). Never use stakes shorter than 30" — they get lost in the foliage.

Business License

A business license is important and usually very simple to obtain. Call the state licensing department for details. Fees vary but are typically under $100.

Soil Test and Soil Amendments

A laboratory soil test and subsequent soil amendments are crucial start-up elements. Find a good laboratory that tests for trace elements, ion exchange capacity, etc., and follow their directions. This could cost in the $200-400 range, depending on the amendments needed for your acre or two.

Summary: *The above is a basic starting package and will vary by location, circumstances, personalities, climate, and so on.*

Giving Tractors Their Due

It would be remiss not to add something positive about tractors. Bear in mind we've already underlined the fact that they're probably not essential at first.

Some believe that a main reason to get into farming, for guys, anyway, especially middle-aged guys, is for the chance to drive a tractor. Well . . . it's true. And it just may be worth all the sacrifices such as lack of health care, retirement benefits, and other perks associated with "real jobs," when a guy gets to wear whatever he chooses whenever he chooses . . . and drives his own tractor. No doubt some women farmers share this reaction — but it's probably not a principal motivating force.

Our farming team considers tractors an important element in some respects. Half of our group is a bit leery about them after she accidentally crunched her foot while learning to operate our early backhoe attachment.

30

A good lawn tractor with trailer can be very handy as a lawn mower, collecting wonderful mulch and compost material from paths, and as a general purpose hauler — especially if it has the booster like John Deere's option to better pick up damp grass.

A larger tractor with front loader, brush cutting device, and other attachments is great for turning compost, digging holes to bury large dead animals, rototilling, chipping and plowing.

A specialized row crop tractor is good for weeding and cultivating in large fields.

Chicken tractor — moveable chicken pen allows the chickens access to delicious (to them) weed seeds.

But none of the above is essential on a two-acre plot. We have, over time, acquired each of the above, including some that have come and gone or worn out. The latter were cheap lawn tractors that self-destructed after two or three years. The menagerie now includes a Sears 14-HP lawn tractor, a 12-HP Kubota diesel with front loader and brush cutter and box scraper, and two old Allis-Chalmers Model G row crop tractors, perhaps rated at about 9 HP. The truth of the matter is that the Model Gs have never been used for anything other than giving rides to the grandchildren and are perpetually for sale.

Lawn mowing, compost turning, driveway maintenance, brush hauling, and similar chores can all be done — much more slowly and laboriously, but quite satisfactorily — by other means. Tractors rarely enter our actual growing areas; we avoid their inherent soil compaction by using wheelbarrows and walk-behind tillers and hand tools.

Conclusion: A good lawn tractor with trailer is a valuable tool in most cases, and a larger machine with a front loader comes in very handy.

So if it is financially feasible, we'd suggest starting with a good lawn tractor and trailer, then consider a larger multipurpose machine after several years of discovering what that machine could do for the operation.

Human tractor — allows guys access to driving a tractor.

The old John Deeres and Ford 8 and 9 series are well known and often recommended as inexpensive power for small farms. That is certainly true where there is hay to be cut and hauled, logs to be skidded, major brush cutting and so on. They are, however, too large and heavy for maneuvering around on a small place, and have little application to intensive row-cropping. The newer small (up to 25-HP) tractors have lots of timesaving features like four-wheel drive and hydrostatic transmissions on lighter, highly maneuverable frames. They are versatile and effective, but expensive, and minimally useful for row-cropping.

Before Buying

Before buying a large tractor, contemplate removing one of its rear wheels to repair/replace the tire, or get a quote from a local mobile tire repair company for the same chore. Our little human-scale Kubota looks really good in that scenario!

Before we had much idea what we were doing, we bought a used small diesel tractor, one that had lots of attachments (including the infamous backhoe). It was useful in some ways but basically was too large to be an appropriate lawn mower, and too small to adequately perform heavier tasks. It was definitely the wrong machine for our situation and had to be resold.

Some may enjoy advertising and selling equipment. To us it's a time-consuming hassle that is better avoided.

If you don't already know this, be aware that it is not fast or easy to switch attachments from backhoe to mower to chipper, and so on, with a three-point hitch. That's one reason so many farms have numerous tractors. Farmers tend to attach accessories semi-permanently and switch tractors instead of accessories.

6
Setting Up the Land

A year of activities gets the land and you ready to grow crops.

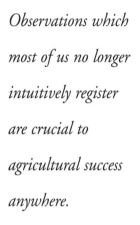

Observations which most of us no longer intuitively register are crucial to agricultural success anywhere.

"If I could start _____ all over again, I'd do it quite differently." Everyone has said that about one thing or another. As we contemplate moving the blueberries for the third time in as many years, we think about how nice it would have been to knowledgeably plan our growing site, and how much time and effort could have been saved.

So here's what we would have done had we known more and had the luxury of a year to get prepared. Follow this sequence if you possibly can.

Test the Soil, then Aerate and Amend

First, we would have completed a soil test with a reputable soils lab, determining what we had in adequate quantity and what amendments were needed. Then we would have hired a "big-tractor neighbor" when the soil was dry in mid- or late spring to subsoil then rototill the entire growing area a year before crops were to be planted. Then in between the subsoiling (ripping — but not turning over — the earth with large teeth to a depth of around two feet) and rototilling we would have spread the recommended amendments on the soil.

Feed the Soil Organisms

Then we would plant a cover crop like buckwheat or annual clover to "feed the soil." That would be plowed under or rototilled lightly to avoid excessive pulverization of the soil when about half the plants bloomed, followed by repeating the cover crop cycle. Three or four successions could be thus plowed in during that growing season; this new organic matter in the soil would encourage all kinds of microbial soil life that would in turn promote healthy growth of our desired crops the following year.

When the last batch was turned under in late summer, we would then plant a winter cover crop of oats and Austrian peas or winter rye to protect the soil surface from pounding winter rains. The pea seeds, by the way, do not germinate well after a second year in storage.

Cover Crop Caution

Notes of caution: if some rye is planted early enough, like in early September in this area, it may grow so well that by springtime it has root masses almost impossible to properly till with any walk-behind rototiller. That happened. Clumps had to be pried out of the ground by hand.

Another problem occurred with one vetch variety: it was so rampant that it quickly got out of control and could not be tilled under. Check for those sorts of risks with knowledgeable people in your local area before you select from the array of available cover crops.

Layout Plan

During the first winter we would have planned and laid out the growing areas on paper. Since our main growing area is roughly 100 by 200 feet we divided it mostly into 10 foot by 100 foot growing zones. Thus we can plan crop rotations precisely on paper and on the ground.

Planning for the Long Term

The winter planning we wished we had done lays the foundation for everything that comes later: where to put the perennials, and why; where to build the fences; where to build the permanent chicken houses; where to run the underground water lines and where to put hose bibs; where to build raised beds; where to lay out paths that accommodate different widths of wheelbarrows, tractors, etc.; where gates will be placed for efficient goings/comings.

Don't skimp on paths. Leave plenty of room to maneuver carts and mowers. Main arterials should be about six feet wide; the narrowest in any case should be three feet.

Closely Observing Your Property

Over the course of the first year one can get a sense of "mini-microclimates;" where is the heaviest frost? Where are warm spots during cool evenings? Where is it wettest longest? Where does rainwater soak in quickly and where does it puddle? What is shaded at different times of day? Where does snow melt first (other than over the septic system)? What are the prevailing winds and where are they most impeded? Least impeded? Where does runoff water go and how does it get there? Which weeds are growing where? Where different weeds grow tells a good bit about the condition of the soil in those locations. See books by Charles Walters on the subject of weeds.

Observing Becomes Automatic

The preceding are key observations which most of us no longer intuitively register since our culture has little use for the information — yet they are crucial to agricultural success anywhere.

Interestingly, those observations do start to become second nature after a few years of attention, just as Bob still automatically notes the nuances of terrain around him as a leftover from his infantry experience in Vietnam thirty years ago.

Acting on Observations

That year of observation would have led us to:

1. Build a tall fence around the entire growing area perimeter — we waited several years and experienced some easily avoidable problems.

2. Lay out all planting areas in linear patterns — we spent a month or more hand digging concentric circular raised beds in one 60 x 60 area — it was beautiful but completely impractical for weed control or watering or getting around in.

3. Locate tall, sun-blocking perennial plants like raspberries on the northern instead of southern side of the area.

Flowering dill.

4. Create wide grass paths around the inside perimeter of the fence so the lawn tractor and trailer could easily maneuver to all parts of the growing area without crossing growing beds; we did properly site the permanent raised beds in some of the wettest areas, thus allowing earlier planting in those areas, uninhibited by muddy ground.

5. Lay out the water lines and faucets so that there would be a freeze-proof faucet every 100 feet along the inside of the fenceline.

6. Build a permanent chicken house, with a freeze-proof water source, accommodating 100 birds, right in the center of the growing area, so, with moveable fencing, the chickens could clean/fertilize/till parts of the area on a rotating basis. We do now accomplish this with moveable fences and mobile shelters; it would be simpler as described.

33

Summary of Preliminary Year Activity

For a new one-acre operation we suggest testing the soil, subsoiling — if locally appropriate (check with the Extension Service) — adding amendments as recommended, tilling, raising a summer's worth of cover crops, installing a water distribution system, planting a winter cover crop, and then fencing that acre. Half of that space will stay in cover crops for the first several years.

Winter is a good time in temperate climates to build raised beds, hoop house temporary shelters and chicken houses. We suggest starting with one hundred linear feet of four-foot-wide wooden-sided (using at least 2 x 8s) raised beds for salad greens. You can add more later.

You may want to select sites and plant a few fruit trees that first winter. Plums are especially good because they grow fast, produce early fruit, and are easy to pick and maintain.

Getting the Chickens Started

Chickens are a crucial part of the operation; starting with 25 might be good, and you can add/delete as you observe how they fit your scheme. They need varying indoor space depending on winter weather. Ours can and do go outside every day of the year, so their floor space needs are less than in cold, snowy climates, but we still give the two layer flocks of about 100 birds each at least one and a half square feet of indoor walk-around space per bird. A third group is purchased as day-olds in late August, and lives in moveable shelters in the growing areas until they start laying (January-February), at which time they move to permanent quarters, replacing the eldest flock.

Chicken Houses

There is a real art to designing an efficient hen house. Check your library for books on the subject. Joel Salatin's book, *Pastured Poultry Profits* and Andy Lee's *Chicken Tractor* provide a great deal of information about innovative housing/pasturing techniques appropriate to a small farm.

There are two things we might add and emphasize. One is the value of a smooth concrete floor in any permanent hen house. That makes it much easier to thoroughly scoop out the straw/manure mix so important to the compost pile. We perform that maneuver about six times a year. The second point is related: always wear a good respirator in the process, because our predecessor on this farm reportedly died from an infection related to inhaling chicken manure dust (in another area of the country).

Other Concurrent Planning

During this physical design and implementation period, there are two other nearly concurrent all-important activities: planning the marketing strategy and developing crop selection/growth plans.

These two factors are the essence of the farm as a business. They are completely intertwined: you can't sell it if you don't grow it, and you won't sell it if it's not popular and presented at the right place at the right price.

You can't sell it if you don't grow it, and you won't sell it if it's not popular and presented at the right place at the right price.

7
Marketing Strategy

Follow the early steps, then your course will become obvious.

Our experiences, both good and bad, suggest the following beginning strategy.

Find the Nearest Good Farmers' Market

Ask your Extension Agent and local people where the nearest good farmers' market is.

Contact That Market's Manager

Contact the market's manager as far in advance as possible to determine your eligibility to sell there, and how they operate. That manager has a wealth of information about what works at that market, what products sell, how they are sold, what the competition is like, and who else to talk to about what you plan to do.

Contact Some of the People Selling at the Market

Ask them what works and what doesn't. Some farmers love to share information. Some don't and won't. Those who will share are probably your very best source of start-up information, though it's best to take such data as a preliminary guideline until corroborated through additional data or experience. Even experienced growers sometimes don't have a grasp of what actually makes money for them and what does not.

Political Red Alert

Another cautionary note should be added here: what people like to call "politics" at farmers' markets is often intense. That is natural when you consider that a market is simply a number of vendors located in close proximity, trying to sell essentially the same products to the same customers. Booth location within the market, who is next to whom, pricing structure, and reselling items purchased elsewhere, are all on the "potentially contentious" list.

Our early experience with a very small local Saturday market was most disappointing — grossing $30 to $100 for a day's selling; later, at a very large metropolitan market it was in the $100-300 range. The third one we tried, in the University District of Seattle, was terrific for organic producers. We took in $200-$400 per Saturday; bigger growers with lots of product and experience in such markets were no doubt in the $1,000-$2,000 range.

Varying Viewpoints About Farmers' Markets

Some growers say they love the camaraderie and don't feel like they are in competition with other farmers at a market. Others describe markets as "snakepits." This points to another truth in operating a small farm: the physical and metaphysical manifestations of the enterprise almost completely reflect the personalities of the farmers!

Set a Longer Term Marketing Goal . . . Sell Ever Closer to Home

Selling at a farmers' market should be considered as your entry point in a marketing plan; you may like it, and choose to focus on it exclusively. Or you may choose, as we did, to use it

Selling at a farmers' market should be considered as your entry point in a marketing plan.

as a one- or two-year springboard to develop some customer loyalty that can then be directed into subscription and/or on-farm sales. Or you can sell at one or more markets and simultaneously sell in other ways . . . there are lots of options and they will unfold as you gain experience.

The Farmstand

Our little farmstand represents a big chunk of our business. It is a simple, clean, attractive 7' x 14' structure along the driveway near the house, about 100' from the five-car gravelled parking area. There are thousands of various such examples all over the country. Look around and find one you like; imitate it at your farm if you have the space and proximity to reasonably well-travelled roads.

The function of the stand is to protect the produce, a cooler, and the customer, from the elements. Ours can display about the equivalent of six apple boxes on a counter, and six more on the floor underneath. There is space for a hanging scale, empty sacks, and a change box for customer convenience. That self-service change box worked pretty well for five years. Then a local out-of-control teenager caused us to go to a locked metal cash box, with a small drop-in slot, firmly attached to the wall.

A farmstand should be easy to access for both customers and farmers.

Our strategy is still evolving, but is built on the idea of drawing in closer to the farm all the time. That translates to the majority of our gross sales from subscriptions and the farmstand, with smaller amounts from florists in Seattle, a local grocery store and restaurant, and one or two Seattle event sales per year. The "closer" we get, the farm becomes more profitable, and we save more time and energy.

We firmly believe the same principle holds true on the national and state levels. Unfortunately, federal/state agriculture policy flies in exactly the opposite direction: "grow more for export" is the theme. But much of our commodity export business is subtly — or directly — subsidized, or even a giveaway, paid for by the American consumer, while now importing more than we export. It's a costly proposition; we should focus on feeding ourselves locally.

Have an ultimate goal of finding profitable market niches that you really enjoy fulfilling.

8
Crop Selection and Planning

"It's like Christmas every week, when we open our subscription bag."
— a subscriber

Having learned all you can of what sells at the market, and what grows well in your area, you are ready to plan the first year's crop selection. This will include what, where, when, and how much.

The "how" to grow will not be discussed since there are excellent books on the subject, including our favorites, *The Organic Method Primer*, by Bargyla and Gylver Rateaver, and *The New Organic Grower*, by Eliot Coleman.

What we grew in 1995, how much growing space it occupied, and how often we sent it out to our subscribers is shown in Appendix B. Here is how to establish your own list.

What to Grow

What to grow? Your clues about crop selection come from the local sources noted in Chapter 7, remembering that you want to differentiate yourself just a little if you can.

Differentiation can be achieved by how you present yourself and your produce rather than just by having *different* produce. People usually come to fresh markets to find fresher food, vegetables that have great-grandma's garden taste, and chemical-free food. Many also enjoy some little connection with farmers. Don't underestimate that latter interest!

We frequently hear from a wide range and status of customers directly or indirectly yearning to be connected with something real, something basic, something of the earth. We are not only growing "things" for them, we are providing a civilization-enhancing service.

Some Specifics

For our area, the following crop choices are popular and command a reasonable per-unit price (in no particular order): peas, pole beans, broccoli, vine ripened tomatoes, Delicata winter squash, cut flowers, fancy head lettuce, basil, fancy salad mix, range eggs, any kind of berry, leeks, sweet onions and elephant garlic. Find out from local growers what crops meet these criteria in your area.

Where to Grow

Where to grow? That is, where within your growing area should each item be?

This is one of the most interesting and complex issues you'll face. As earlier stated, we had nearly forty years of successful technical and general management experience between us when we took up farming, and now realize that diversified farming is by far the biggest — but most fun — management challenge we've ever seen.

Eliot Coleman's book listed above provides a detailed description of how to choose what best goes beside what and what best follows what in a crop rotation scheme. We used his guidelines and developed a comprehensive plan, frequently modified as we gained experience, and recommend you do the same. It is not at all an exact science.

Diversified farming is by far the biggest — but most fun — management challenge we've ever seen.

How Much to Grow

We did not find adequate guidance in any source to help decide what quantities of each item to grow. Trial and error is slowly providing that education, and here are some gleanings from coping with the "How much?" question.

Some items are limited by the amount you think you can sell, others by the labor involved in planting, maintaining and harvesting. Backwards planning, as refined in assault ship loading plans for the World War II Normandy invasion, is the intellectual place to start working this out for your particular situation.

Planners decided what the combat troops would need in the first hours, and loaded that last onto the ship so it could be first off. They continued that line of thinking so everything in the load was prioritized by how soon it was needed, and then loaded in reverse order.

Our corollary is: How many boxes of what item do you want to take to a farmers' market, by week, next season? That's the defining issue.

When Can It be Ready?

A little side trip into the "when will it be ready?" estimate is needed before the actual planting amounts can be determined.

Added Complexity

We diverse growers add major new wrinkles to backwards planning. We don't just prepare for one event; we have each week of a growing season to consider as an event, and that means further careful planning related to "succession planting" — the practice of planting at intervals so the produce comes along in a stream, rather than all at the same time. And some crops, peas for example, properly grow only during fairly brief parts of the season, while others last most of the warm months, so that aspect must be considered. These issues are the key to success with a CSA farm.

The good news is that this crucial time planning is easy to figure out with a calendar and some seed catalogues.

Start by circling on a large calendar the various dates when everything you want to grow can be transplanted outside or direct seeded outside. Such information is found in the better regionalized seed catalogs, like from Territorial Seeds in our part of the country.

The catalog or other local sourcebook will also tell how many days it will be from that point to date of harvest, and how long the harvest may continue. So you can tell approximately — they always use optimistic estimates in seed catalogs — when you will have each item for sale, and, counting backward on the calendar, you will see when transplantable items must be started in flats, and when others should be sown in the garden.

Those planting dates should then be duly noted on the calendar — a very large calendar, to accommodate all the information you will write thereon.

Hypothetical example: If your chosen variety of spinach can be direct seeded from March through August, and it matures in 40 days, you know that in theory you may plant spinach seeds every Tuesday from March 1st through August 31st, and have bunches of spinach available every week from about April 10th to October 10th.

That is the theory. What keeps everything so interesting is, that, in practice, seeds germinate more slowly in the early season cold ground and grow on faster at certain temperatures, then slow down again at certain higher temperatures, so you have to adjust for those factors based on experience.

How many boxes of what item do you want to take to a farmers' market, by week, next season?

Quantities

Now that you have "guesstimated" what will be available when, you may then decide how much is wanted of each item during its particular production timespan.

If you decide that 25 heads of broccoli per week is what you would like to take to the market, and your earlier work from the seed catalog indicates that broccoli can be harvested all during the month of June, you have the basis for your broccoli growing plan.

Starting Plants

We firmly believe in starting everything possible in flats in a covered growing area.

Transplanting those starts later on may be labor-intensive and time-consuming and wearying but well worth the effort. That process sets the stage for producing exceptional produce.

We start everything indoors except potatoes and carrots, usually in 200-plug flats, and plant 25% more plugs than the number of plants we calculate we'll need. Then we can choose the best seedlings to plant outside and discard the rest.

Loss and Quality Control Factors

For the broccoli example, we would begin by estimating that to get 25 good heads per week we would want a safety margin of an extra ten or so growing plants, thus allowing for various potential quality problems between transplanting out in the field and proper maturity. That means we would plan field space for 35 plants to be planted out from transplants each week.

Since we use a 25% safety factor between the seed and the transplanting time, we would plant 35 plus 25%, or 44 seeds, each designated week in April or May, as determined from the seed catalogue, to provide at least 25 good heads each week in June.

A beneficial bee pays a visit to the flower garden.

All other vegetables can be determined the same way. Berries and fruit are a little harder to estimate since they are influenced by longer term factors, but their productivity can be approached the same way.

Self-Imposed Production Limits

We have established personal limits on several popular items. There is more demand than we choose to fill, but because they require too much time we limit output as follows:

Pole Beans: 200 row feet is all we two can (barely) keep up with during their long season.

Peas: Same, even for their short season.

Raspberries: About 100 row feet is our limit.

Tomatoes: 250 plants if they have to be trellised.

Corn: 500 stalks, mostly because we shuck and carefully check each picked ear for borers.

Potatoes: About 300 row feet because digging them is hard on shoulder joints.

Other Limitations

There are other general production limits based on our sales experience:

Herbs: Are attractive additions to a farm stand but sell very poorly. Basil is an exception. We told farm market customers (while we did that form of marketing) what we had in the herb line — a lot — so they could special order, and some things like Sorrel and Rosemary did develop a small sales niche. Our subscription customers appreciate small quantities of the major herbs from time to time, but herbs are not used consistently by very many people, even with weekly helpful recipes and other promptings.

Cucumbers: Sell well at the farmstand for the first weeks of production, then drop off to a steady but smaller number.

Summer Squash: Is the same as cucumbers.

Kohlrabi and Bulb Fennel: Are interesting but sell sparingly.

Cabbage: Is not a locally popular raw material in this era of prepared foods. It may be much more important elsewhere, especially for certain ethnic groups.

Rhubarb: Is not well known by the general public . . . we thought it was widely known and admired, but its followers are mainly the old-fashioned pie bakers, of whom there are far fewer than there used to be! It's hard to beat a good strawberry-rhubarb pie.

The Next Step

With all the above information and backward planning, you can determine what you will grow, when to start it, when it will be ready, and how much to plant.

Space Allocation

That leads you to how much overall space is required for each item, by following spacing guidelines in the seed catalogs or other sources. We suggest starting with those guidelines and then developing your own scheme. Here are some concepts we follow:

For Dense Planting

Using good homemade compost and building your soil allows much closer plantings than those shown in the seed catalogs. We grow lettuce and the smaller leafy greens about four inches apart in rows four inches apart, in four-foot-wide beds; kale and larger leafy plants do well at sixteen inches in all directions.

Squash blossoms.

For Wide Apart Planting

At the other end of the plant size spectrum, we go to greater than recommended intervals for plants that sprawl and need frequent picking, like summer squash, cucumbers, and tomatillos.

Next season we will put all such crops at three or four foot intervals, with six feet between rows to facilitate passage as well as use of six-foot ground cloth for weed and moisture control. This worked very well where we could afford to use the rather expensive ground cloth in past years. The technique uses more space but is a time and water saver. Usually time and water are much more of a premium concern than is space, even in a small growing area.

And for Standard Spacing

Sprawling plants that are only picked once, like winter squash and potatoes, are planted at catalog-recommended intervals.

More Space Layout Planning

With the data developed to this point, you can determine how many total square feet of space are needed for what you totally want to grow, including 15-20% of total space for cover cropping to rest and enrich the soil. Calculate how many square feet of actual growing area you have, deducting paths and nonproductive spots like around faucets.

You may have planned crop for more space than is available, and must scale back some items, or your growing needs may not fill available space; in the latter case, fill space by leaving it in "green manure" cover crop all season long or rotate chickens through it in 4' x 8' moveable pens ("chicken tractors") or in a temporary shelter surrounded by portable fencing.

Putting It All Together

The next step is to literally create a jigsaw puzzle so all the pieces can be fitted together into a cohesive visual overview of the growing area. Again, Eliot Coleman's *The New Organic Grower* describes this process in detail.

We used 3 x 5 cards to visually represent one thousand square feet of desired crop; thus we wrote "tomatoes" on three cards since the number of plants we wanted to grow were calculated to fill three thousand square feet. Other 3 x 5 cards were cut in half for crops needing only five hundred square feet, and so on. Eventually, there was an annotated card (or more) or portion of a card pictorially/quantitatively representing each of the 40 crops we wanted to grow.

It isn't necessary to establish a card for each variety within a crop family; thus the three tomato cards represented the cumulative space required for all twelve varieties.

Needless to say, forty crops required quite a few cards in total.

Juggle the Pieces

One can then move the cards around on a tabletop, grouping them according to what goes best next to what and what best follows what in the rotation plan. This takes some real maneuvering and trade-offs. But it is a tremendous learning experience and raises issues that encourage further reading, observation and study.

Rotation Plan

Our block of tomatoes, at three thousand square feet, turned out to be the biggest single item for the 21,000 square foot "most sunny microclimate" portion of the growing area. Thus everything else that wanted maximum sun had to be fitted into internally agreeable 3,000-square-foot blocks that will follow each other around over the years.

There was a lot of juggling and compromises and guessing that eventually fitted all those crops into a scheme of seven such blocks . . . block A moving next year to where block B is this year, block B going to where block C was, and so on, thus each block will take seven years to return to its original spot.

For the other geographical area of the main field, the 1,000 square foot "sometimes shaded" part, we followed the same procedure to fit in crops that appreciate less sun, and came up with four blocks of about 250 square feet each. So the entire main growing area is divided into two rotation groups — one seven-year cycle for those who always want full sun, and a four-year cycle for those that can handle partial shade. The south field is divided into four blocks for plants that require little attention.

Six or more years between repeats on the same growing plot is probably optimal for each group, but creating six blocks in the "less sunny" microclimates and the south field didn't seem to work with what we wanted to grow. Compromises are often unavoidable.

We allocate 15 to 20% of total space for cover cropping to rest and enrich the soil.

RESULTS FOR 1996

Main Field — Full Sun (21,000 sq. ft.)

Zone#	Crops
1, 2, 21	Flowers, Peppers, Eggplant, Cover Crop
3, 4, 5	Winter Squash
6, 7, 8	Leeks, Shallots, Onions, Cover Crop
9, 10, 11	Tomatoes
12, 13, 14	Kohlrabi, Beets, Spinach, Chard, Cover Crop, Pole Beans
15, 16, 17	Cover Crop, Corn, Cukes, Melons
18, 19, 20	Broccoli, Summer Squash, New Zealand Spinach, Tomatillos, Cover Crop

Main Field — Partial Shade (1,000 sq. ft.)

Zone#	Crops
1	Onions, Garlic, Cover Crop
2	Salad Greens, Cover Crop
3	Cover Crop, Salad Greens
4	Salad Greens, Cover Crop, Carrots

Swiss chard.

South Field — Full Sun (10,000 sq. ft.)

Zone#	Crops
1	Shelling Beans
2	Potatoes
3	Winter Squash, Pumpkins
4	Cover Crop, Grain

9
Starting Plants

Almost everything grows better when started indoors.

At several points in earlier chapters we have mentioned starting plants. There are several ways to do that, most notably direct seeding in the ground where they are to grow, or seeding in flats for protected start-up indoors — followed by later transplanting outdoors.

There are many successful techniques for planting either way. Different people have different views about which general approach is best. Our choice is seeding in flats indoors. The only items we direct seed outside are carrots and potatoes, and sometimes peas.

Planting methodology rationale . . . sounds like a bureaucratic term . . . but is the general heading that comes to mind concerning planting choices. Some like the simple, fast, satisfying qualities of direct seeding. We prefer the "known quantity" and precision of starting plants indoors, then transplanting outside later. This latter method is costlier and much more time-consuming in all phases. But we believe its virtues outweigh its faults.

Advantages of Indoor Plant Starts for Small Acreage

The advantages of direct seeding were noted above. It is a fast, easy and economical procedure; so why not love it? Local climate somewhat legislates against it in our case.

By seeding plants indoors we get a big jump on the warm season. Our soil doesn't really warm up until June, so seeds of warm weather crops like tomatoes and squash and corn often rot in the ground if planted before early June. If those plants are started indoors in March/April they can be set out as vigorous, ready-to-go plants in mid-May — they won't grow very quickly, but they won't rot, and they do survive and grow slowly until the big takeoff in late June or early July.

A second advantage is that you should always start more plants than you need, so usually there are high quality extras to sell at the farmstand or wherever. They can be good money-makers early in the season before much else is saleable.

Another main advantage has to do with known quantities: knowing what you have at all times.

Sometimes seeds don't germinate well. We like to know that early on, so there is time to adjust by getting more seed from another source and replanting, without waiting into the season to find out there is a problem. One can establish ideal temperature, moisture, and soil conditions for little seedlings in flats or small pots, and if they don't germinate within the designated time, you know there is a problem with the seed.

You also know exactly how many survive to the "sturdy seedling ready for transplanting" phase, and, since you started more than you need, you can plant out the very healthiest ones in rows of sturdy little plants.

They have a nice jump on the weeds and pests, and no thinning or replacing should be required, which is much different than in direct-seeded areas where the weeds and crop seeds come up together. Actually, since "weeds" are just other locally well-acclimated plants of other species, they won't truly come up together; the weeds will come up first and probably strongest. That means a lot of hand weeding after they are both up to the point where one can tell them apart.

Sell your extra plant starts — they are good money-makers early in the season.

Carrots are especially difficult in that regard. There are techniques with weed-killing propane flamers that apparently work well with carrots. We haven't tried that yet.

Good-sized seedlings are generally too big to be wiped out by crows and pheasants. That solves one of our big problems.

The sturdiest seedlings grow into the best mature plants and produce the best crops. Some of us like to cheer (root?) for the weak little underdogs . . . but grudgingly admit the truth that the sturdiest produce best.

Plant Start Materials Required

Earlier we said that you need a good-sized temporary structure to shelter some items and some processes. That mostly meant the activity and supplies associated with starting plants.

For a physical layout, one needs a table space about six feet long and two or three feet wide, at a comfortable standing work height, and within close proximity to the planting supplies storage area.

Heating cables are necessary for most early seed starting, so a 4' x 8' table is a good way to accomplish that (more details to follow). After removal from the heating cable table, it is best to let starting flats sit on hardware cloth or wire fencing material stretched over a sturdy wooden frame to make access more difficult for slugs and sowbugs, and to avoid letting aggressive roots go from the flats into soil of any sort.

Greenhouse, chicken house and tool shed combination.

Necessary start-up supplies include sterile growing medium, enough flats with plastic dome covers to start the number of plants you earlier planned, flat inserts of appropriate sizes, and some soil amendments. Let's consider each of those categories.

Growing Medium

First, there is a distinction between a "starting" medium and a "potting" medium. The essence is that a starter is fine-grained and has no, or very little, food for the plant — the initial food is self-contained in the seed — while a commercial potting mix is larger textured and has fertilizer to help the plant grow on past the seedling stage.

We start plants in a sterile starter medium, then add a mix of solid and liquid organic fertilizers as they start to grow.

There are many starting/potting brands on the market. Only a few of the mixes are OK for use by a certified organic grower since most have some quantity of chemical fertilizer in them, and some are downright "forest product junk" heavy on bark bits and pieces of logging truck tires. (Well, maybe it only appears so.)

Check them out by appearance and feel. You will have to find out fertilizer content from the manufacturer; it's usually not listed on the bag. Whichever brand you choose, try a small bag very early in the season before you buy your whole desired quantity.

You can make your own starting/potting soil fairly easily, using peatmoss, vermiculite, perlite, and additives as desired. We think it's worth buying a good commercial batch for the sake of convenience. Peaceful Valley Farm Supply has a good one. See Appendix A. We are eagerly awaiting the development of an organically acceptable wetting agent to aid in the speedy initial saturation of flats.

Planting Flats

The flats we use are 10" x 20" and will hold inserts of various configurations. Several sizes of seedling inserts are required: one size for starting small seeds (most veggies and flowers). We prefer one with 200 plug cavities, and at least one larger size version for squash and pumpkins that have larger seeds and require more space to grow.

In the case of tomatoes and peppers that must be started two months before transplanting outside, repotting from the 200-plug flat into four-inch or larger pots that will allow the plants to gain substantial size before setting out in the ground is a necessity. Most other plants do not require repotting. They can be transplanted directly from the 200-plug flat into the ground.

It is a fact that the pot size absolutely limits the ultimate size of the plant therein.

Anyone who has ever bought potted plants to take home for replanting in the garden has seen the typical "rootbound" situation where the roots are a dense, intertwined mass. That is bad for the plant. Roots should just be reaching the walls of the pot at the time they are ready for planting out; if you are not able to plant them out at that development stage, they should be repotted into a larger pot in the interim.

Thus, if you start plants like tomatoes and peppers and other heat-lovers in late February, for transplanting in May or early June, you will repot one or more times after the initial seeding in a flat or plug tray.

Amendments

As organic growers, we use materials like kelp meal; bone meal; meals high in nitrogen, such as blood or alfalfa meal; rock powders; agricultural lime, and compost. We avoid the use of cottonseed meal due to the concentration of toxic chemicals used by cotton growers.

Most manufacturers of the starting medium add lime to their mixes to ensure the medium has a neutral pH factor, balancing the acidic peat moss. Because the seed contains all the nutrients it needs during the germination stage, we do not add anything to our starting mix. However, as soon as the first true leaves develop, we begin a weekly application of a weak solution of foliar fish and seaweed. The addition of any other amendments is done both during the garden soil preparation stage and at the time of transplanting.

Magnesium Overdose

Even organic amendments can be overdone, or wrongly done! Early on, we used a lot of lime on the growing field because our soils are basically acidic and calcium is important to make the nutrients in the soil available to the plant. What we didn't know at the time is that there is a difference in types of lime and the common type sold at our regional mass merchandisers is dolomitic lime, with a big component of magnesium, which our soil doesn't need.

Our magnesium overdose was only discovered when we had a professional lab thoroughly test our soil. Prior to that, we had done our own soil testing with one of those inexpensive little kits that gives some indication of pH, and nitrogen, phosphorus, and potassium levels. We were told by the lab to use only agricultural lime, not dolomite lime. Our magnesium content was at a level nearly toxic for some vegetable plants.

So why do all the local stores continue to sell only dolomite lime? We discovered that agricultural lime is available only through agricultural suppliers.

Other Aids

Plastic domes that fit over a 10" x 20" flat are helpful during germination. Bottom heat is necessary in the early season to insure that most types of seeds germinate. The catalog from Johnny's Selected Seeds is an excellent reference for optimal seed germination temperatures. Heating mats are quite expensive, but a good setup can be fabricated at home by buying a

Even organic amendments can be overdone, or wrongly done!

45

heating cable kit, then looping the cable back and forth on a piece of plywood covered with a thin layer of sand. Our plywood rests on sawhorses about kitchen counter height. The flats can sit directly on the cable in the sand. We've had good luck with that arrangement, spacing the cable at just over two-inch intervals. The cables get hot enough to partially melt the plastic flats if the cables are too close together.

Picking green beans.

One heated 4' x 8' starting table is adequate for 19 flats at a time — that's 3,800 plants. It usually takes just a few days to germinate a batch, then remove the domes and take the flats off the heat but keep them inside. Over the course of the year we start a total of 17,000 to 20,000 plants; roughly half occur in that early period when bottom heat is necessary.

There are many variables associated with planting seeds. Some do well only if buried, some only if left on the surface; some want to be cold, some want heat; some have to be frozen for a period of time or scratched with a knife. The seed catalogs, with the exception of Johnny's, are often overly general or vague about these factors. Two excellent books are Nancy Bubel's *The Seed Starter's Handbook* and *Park's Success With Seeds,* by Ann Reilly, Park Seed Co.

Transplanting

There is not a great deal to add on this subject. We sprinkle a mix of compost, kelp meal, bone meal, rock dust, and alfalfa meal in every planting hole, and when there is an overabundance of chicken eggs, we add one of those for any of the larger plants. It's not clear how much good the eggs do, but we like the concept! It was a great way to get rid of over 1,000 tiny pullet eggs no one seemed to want to buy last spring.

One technique that does work well is to plant tomato plants on their side in a shallow trench. Only the top several inches is bent up and left above ground. The buried stem and branches quickly develop roots that enhance overall growth.

10
Operating a Subscription System Farm

"These are my farmers," said our subscriber to her friend.

We have talked about all the preliminaries and prerequisites. Now on to the details of the subscription system that is the core of the operation. That core may change over the years, but in the midterm it establishes stability and predictability and connection with the ultimate customer that is the crucial link in small-scale agriculture.

Let us assume you have been through the start-up activity and have two or more years of experience selling at a farmer's market. You will have some idea of how you do things, what you do best, what seems to grow best, and, in general, if you want to be serious about making a living on a small farm. If you have decided that this life is for you, then the next step is to develop your own medium-term farm plan.

A Mid-Range Plan

Everything evolves. Your notions about what you want to grow, how to lay out the farm, and how you like to sell may change substantially over time. It is painful to think of the numerous "permanent" fencing changes we've made, for example, and even the "permanent" raised bed structures achieved with so much hand digging that have been levelled later. As mentioned earlier, the poor blueberries are in their third "permanent" home in as many years. They are set back with every move; so are our blueberry pancakes.

There is an issue at the other end of the "permanence" spectrum, and that concerns trees. They aren't so moveable after several years; you must make some genuine long-term orchard/plantation plans early on, before you are probably ready, or you will seriously delay arrival of the fruits of those trees. Read one of Bill Mollison's "Permaculture" books before making those decisions.

Knowing that things will change over time, develop a mid-range plan. One part of the plan will concern the physical layout considered in Chapter 4. Another will be the marketing plan of Chapter 5. A subsection of the marketing plan will be how many subscription customers you feel comfortable starting with, and how you plan to find them.

Get two successful growing seasons under your belt before starting a subscription operation.

The Basic Subscription Start-Up Questions

How many subscribers should we attempt the first year?

That varies by individual preference, experience and confidence. We know one two-person, part-time operation that started with only six customers; another took forty. Both survived and both expanded. We also know of failures, where the advance payments had to be refunded in mid-season because the farmer could not handle the demands of the business.

We believe twelve to twenty is a good starting point for those with minimal experience.

Should I begin with fewer customers than I really think I can grow for?

Sure. You can sell surpluses at your farmer's market or at a farmstand on site. The issue usually is not how much you can grow, it is growing balanced amounts of a wide variety of crops to interestingly fill each bag each week.

What are the main problems the grower encounters with this system?

Growing so many different things for a prolonged period of time . . . which requires such good planning and timely execution . . . is both a curse and a blessing to the grower.

The biggest complaint we hear about our system, or those of other subscription farms, is that there is too much of one thing for too long a time. The extreme case was reported by one who claims his farmer provided six bunches of radishes, and very little else, for three weeks in a row at the beginning of the season! Needless to say, that relationship did not endure.

We try to delay start of the weekly pickups into mid-May — even though the pressure mounts to get going earlier — to have an assured supply of diverse items. In fact, we use some items from the previous year, like nuts and dried beans, plus potted vegetables or herbs or flowers for customers to plant at home, to add variety in the early weeks of the season.

The other main problem from the grower's perspective is that this is a season-long commitment. It is a good idea to schedule a one-week hiatus during the middle of the summer. That is OK from the customers' perspective and a nice break for the farmers.

What is the best way to advertise for customers?

Word of mouth is the very best. Our first customers were friends or friends of friends, and only a few of the first twenty came from the various newspaper ads we ran. Tell your friends what you are about; prepare a good brochure and give copies to those friends; take copies to the farmer's market and hand them to customers you have gotten to know a bit. The local newspapers and other media may be interested in publicizing your program if it is somewhat novel in the area. There are various support organizations listed in Appendix D that have brochures and ideas about how to work with the media. Develop a good logo early on and use it on a farm sign; visual advertising will reinforce your brochure and your own sense of professionalism.

Should the farmer deliver produce directly, or deliver to pickup points, or have people come to the farm every week?

In most cases, it is simplest for the farmer to have customers come to the farm. That way they can get to know the farmer and farm premises better, may want to purchase additional items if they see them for sale, and it saves the time and trouble of loading/delivering.

There is an implication herein that the farm is neat, attractively maintained, is easily accessible along decent roads, has parking and a place to properly store the full produce bags or boxes on pickup day.

Some farmers prefer drop-off points at houses or businesses in town, where, in exchange for a share of free weekly produce, the residents will provide secure storage and availability so the other customers can pick up their bags or boxes without travelling out to the farm.

Many of the larger operations have one or more coordinators who set up an entire pickup system, collect the money, and recruit new members. The coordinators would, of course, be remunerated for their services. That reportedly works for many farmers. We prefer not to do it that way because it gets back to the idea, however benign in this situation, of a middleman between the grower and the eater. It is also an extra cost to the farmer.

Community Supported Agriculture calls for direct linkage between the person who grows the food and the person who eats the food.

48

Another option is for the farmer to deliver direct to each customer each week, for an extra charge. We found that to be very popular but it took an inordinate amount of time and added much wear and tear on the van. It may work well in some circumstances, however.

How much should we charge for a subscription?

This may be the biggest question of them all. And it has the most varied answers.

What is so dicey about the whole issue is that no one knows in advance exactly what will go out each week, and therefore what the value in wholesale or retail terms may be for the season. And every CSA/subscription farm sends out decidedly different quantities, different produce, and often prepares and packs it quite differently.

Some develop an annual farm budget, allocate proportionate amounts of costs and gross income to the subscription program, and then divide that portion by the number of subscribers.

Example: If the farmer determines from a well-prepared budget that serving 50 subscribers for 20 weeks requires $20,000 gross from the subscription portion to meet farm financial objectives, each subscription would cost $400.

Some go by total poundage. They plan to provide a certain number of pounds of produce per week to each customer. That seems a strange approach since a pound of salad mix is so much more valuable and time-consuming to prepare than a pound of winter squash, for example.

Our choice was to target the retail grocery store value of what seemed would be a typical bag we could fill in mid-season, then back off from that to give the subscriber some saving versus shopping at the local grocery store. That came out to about $15 per week the first year; we have increased prices about 4% each year based on higher costs of farm inputs, especially seeds and chicken feed. We realize, and mention to our subscribers, that the early season bags will contain less value, but we'll make up for those later in the year.

Bonnie bagging the subscription shares in the prep area.

With our system, we have a target value to achieve each week. That simplifies things a bit. We keep an eye on grocery store organic produce prices, and then fill the sacks to reach our desired value each week, adjusting for past underages as necessary.

Notice we use bags. Most use boxes. Our experience suggested it is far easier to obtain, store and handle used grocery bags than it is to work with more bulky boxes. Also, we feel it is essential to keep most vine-ripened vegetables and fruits in cold storage until in the hands of the consumer, so cold storage space becomes critical. Our old beverage cooler makes a nice cold storage unit for customer pickup, but it will handle only a few boxes versus a maximum of 36 full grocery bags.

As a general rule for establishing seasonal pricing, look at what others are charging in your area. The least we know about, anywhere, is around $12 per week, and the most is around $25. Again, there is no consistency among weekly outputs of any of these, so it is impossible to directly compare them.

"I started out at a very low subscription price because I was new and didn't know if this whole thing would work… now I realize I gave them about twice what they paid for, and for this next season must either cut way back on what they are used to getting, or substantially raise prices to make ends meet, but am very afraid I'll alienate my existing customers." This is a nearly verbatim comment from various new farmers we know or know about. Beware. Be scrupulously fair to everyone, including self, from the beginning.

How does the subscription program affect a typical farm weekly schedule?

There's nothing unique about scheduling in this system. Any marketing method requires similar amounts of work and routine, including deadlines.

We have always spread the customer pickups/deliveries over two nonadjacent days. Thus if 18 are scheduled to get their sacks at 3 p.m. Wednesday, and 17 on Friday, we follow this routine:

Monday a.m.: Pick and wash 20 pounds of salad mix.

Tuesday a.m.: Same as Monday.

Wednesday: Pick and pack all other items for 18 subscribers, in the bags by 3 p.m.

Thursday: Pick salad mix for weekend sales.

Friday: Same as Wednesday.

That is the part that directly relates to subscriptions. In addition, we consider and harvest for farmstand and grocery store sales.

We pick salad again later in the week as needed for sale at the farmstand, and do the same with other produce as it ripens. Everything is picked fully vine-ripened and ready to eat to insure best flavor and highest quality for the farmstand. Grocery stores, on the other hand, want produce that is not yet ripe so it can last longer on the shelf and withstand more handling.

The rest of our schedule is quite variable. It depends on the time of the season and relates to all of the rest of the weeding, seeding, transplanting, watering, etc. Nothing unusual there.

A confession: everyone assumes farmers start work at the crack of dawn. Most farmers apparently do, and relish those early hours. Well . . . we don't arise until after a half hour of classical music and the 7 a.m. news on the radio, including those always-gratifying traffic reports. Our systems simply work more happily with that routine. We do work until dark. And it is a seven day week, with Sunday being a more casual time of doing odd jobs and lesser things, including what we sometimes refer to as "recreational weeding" — weeding an area that is not crucial but will make us feel better when it looks better.

So there is hope for even the early-morning-impaired.

What are the details about preparing a good salad mix?

Salad sales are crucial and rewarding to us. Capitalize on this if you can. Plant and use any salad ingredients you can think of, all year long. You'll get good at planning, planting, picking, washing, mixing and bagging a salad mix. Strive for a taste combination you enjoy.

Bob can pick about 2 pounds of bite-sized greens per hour; Bonnie is faster than that. We soak 4 to 5 pounds of greens, right after picking, in a washtub of cold water for about 5 minutes, then repeat that in another wash tub, watching carefully for bugs. Then after some gentle swishing of the water, the greens are put in plastic laundry baskets for about 20 minutes to drip dry (more than 20 minutes during cooler weather). Then they are bagged in large plastic shopping bags and put in the cooler.

Be scrupulously fair to everyone, including self, from the beginning.

50

After all the greens have cooled a bit, we pour the whole works — up to 35 pounds — in a very clean old bathtub, where they are gently but thoroughly mixed, then we weigh the mix out into half-pound lots and put those into 1-gallon Ziploc freezer bags. The salad mix will retain its freshness well over a week when stored that way in a refrigerator.

How many people will one subscription share feed each week?

That is an utterly unanswerable question, though most CSA brochures include some comment about one share being enough for three people for a week, or whatever.

Bob's 21-year-old son could probably consume one of our bags in two sittings. Alternately, some families of four might dabble at this assortment for a week. For reference, a typical 1995 July bag contained one-half pound of gourmet salad mix, one-half dozen eggs, a huge sweet onion, four or five small summer squash, several kohlrabi, four globe artichokes, a head of lettuce, a big bunch of New Zealand Spinach, and a pint of sweet cherries.

Other CSA farms often concentrate more on bulk and do less prep work — everything we put in the bag is clean and ready to eat, looking just like it came from a grocery store. Some farmers pick directly into the customers' boxes, dirt and all. Their weekly fare might include several pounds of potatoes, several pounds of beets, several pounds of carrots, heads of lettuce, chard, and so on, perhaps up to 20 or more pounds. That might indeed provide plenty of certain kinds of vegetables to feed a four person family for a week.

Fortunately for all of us, there is no single answer. Some folks prefer lots of pounds of unprepped vegetables, and others prefer a broader mix of prepared foods. There is a huge market for each of those styles plus many more!

Is a weekly newsletter important to the CSA customer/subscriber?

Indeed it is. Our customers have said many times that they like to know what's going on at the farm and with the farmers, and they especially like to have recipes for what's in the bag.

Bonnie collects recipes all winter, so we have a repertoire ready to print as different items ripen.

Some farmers consider a newsletter an insignificant, unpleasant chore; we think of it as an important, integral part of our farm program. It usually requires 30 minutes to prepare the night before customer pickup day.

The greenhouse being used as a ripening room for tomatoes that had to be picked early due to cold weather.

Is it important to keep good records; and is a computer essential?

It is definitely important to keep good records. It goes without saying that customer payment schedules and records must be precise; other things like planting dates, harvest dates, quantities planted and picked, and costs are very valuable for future planning.

Perhaps less obvious, but equally important, is that keeping good records constantly reminds us that this farm is our job, not a hobby. It enhances the professionalism needed to be successful.

As for computers . . . we've kept financial records both by hand and on the computer. It seems like the computer takes more time and effort in the long run, even for very detailed records, unless using the computer is second nature.

The concepts associated with the Dominguez/Robin book *Your Money or Your Life* have been extremely important to us in this overall issue of records and tracking expenses. Anyone wanting a simpler, more basic and meaningful lifestyle will find immense value in this book. It has greatly aided us in reaching a debt-free farm situation, plus a savings program that is working.

When should subscribers pay?

There are once again many approaches to this. We want the bookkeeping essentially out of the way before the rush of the season overtakes us; our subscribers pay a deposit in December, then equal installments by the first of April, May and June.

What subscriber turnover rate should be expected?

Some farms report up to 60% turnover per year. Our experience has been in the 10-30% range. One must watch this closely and use year-end anonymous evaluation forms to determine what is causing turnover. There are plenty of reasons over which the farmer has no influence (moving, finally started own garden, etc.) but dissatisfactions must be discovered and addressed if one is to maintain a successful enterprise.

We realize this system is not for everyone and some will drop out every year. But most subscription farms have nice waiting lists filled out by the end of each season.

A totem pole in a wooden wheelbarrow – there's no telling what you'll find around an old farm.

"During the winter I suffer from light deprivation and salad deprivation. There's no substitute for your salad mix!" — a subscriber

11
What's Next?

Refinement of techniques will forever be the quest.

Refinement of techniques is another factor that makes this business so fascinating. Farmers in Asia had achieved a more or less steady state of full sustainability after thousands of years of refinement. Nothing much changed each season. Now we have periodic new technology that can be applied here and there, and widespread instant communications that stimulate the thought process.

Each New Practitioner of the Organic Growing Art Will Add Something Unique

We have done a number of pretty clever little things. Most are not readily explainable in print. A lot have to do with "making do" with materials on hand to solve problems, minor little triumphs that don't amount to much, yet give us considerable satisfaction. You will do that, too.

Our next challenges have to do with considering taking on an apprentice; creatively raising much more chicken feed without putting in the necessary facilities for storing grain; saving seeds from open-pollinated plants; doing what is necessary to retain water in the two runoff ponds, then creating habitat to raise crayfish and fish; replanting the over-age nut orchards; creating better windbreaks; using the land around the ponds in a more productive manner; building an innovative chicken housing complex that will allow easier, more direct access to the main crop-growing area; better insulating the cold-storage room; collecting and using rainwater; integrating a few pigs into the operation; and . . . sounds like a lifetime or two of interesting challenges!

The road ahead will be full of fascinating challenges for you.

12
Other Things to Talk About

Self-sufficiency, partnership, personal power and the big picture.

This book has covered a lot of practical information. There are additional important areas that aren't so directly nuts and bolts.

How About a GI Bill for Small Family Farmers?

It's been said that the most cost-effective federal program ever was the GI Bill that helped educate and buy homes for WWII veterans.

Why couldn't we offer the same concept for those who want to start farms? Educate and encourage people to lease, borrow or rent land for several years and farm as outlined in this book or others. Reaching a target sales level — say $20,000 per year — the farmer would validate seriousness; then the "Small Farm Bill" would offer a no-interest long-term secured loan so the farmer could afford to purchase his/her own farm.

The value to our society would be immense: thousands of new small farms all over the country would create jobs, improve quality of food and nutrition, protect open space, rebuild the countryside, and revive a way of life that enhances the values crucial to our culture.

The loss of the small family farm has seriously damaged our whole culture over the past 60 years. But now, mixed-crop farms as described in this book, and pasture-based small livestock operations described in Joel Salatin's books, offer economically viable options to start healing that damage.

Some Thoughts About the Concept of Self-Sufficiency

Like almost anything, this has a positive and a negative aspect. The positive is a wonderful, though partially illusory, feeling that I and mine can take care of ourselves and control most parts of our daily living independent of what the rest of the world may do.

The negative aspect is the extreme of shutting oneself off from others, the fearful self-obsession that denigrates other people and other ideas, in the name of personal survival. We think it's worth mentioning this because the term "self-sufficient" comes up so often in those colorful publications promoting the joys of rural life.

A Driving Force

This overall notion of self-sufficiency is undoubtedly a subtly potent driving force for many of those drawn to the small farming concept. And understandably so! Several years ago, as we were kindly hosted by Bob's eldest son at his pleasant brownstone apartment in old Boston, Bob had difficulty going to sleep thinking about how we were totally dependent on antiquated electrical, water, and sewer facilities. If a large disaster were to occur — we think about this sort of thing, living in major earthquake country — there would be no water or access to it, there would be no heat or access to it, there were very limited places to even dig latrine facilities, and within several days no one would have food.

City folks, and even most suburban dwellers, are totally at the mercy of systems completely beyond their control. That has to register at some inner level and to some degree impact most people, whether they recognize it or not.

When one grows much of his/her food . . . there's a sense that almost anything can be accomplished if approached in a methodical, thoughtful way.

54

Material Self-Sufficiency

Obtaining a good, healthy dose of material self-sufficiency is usually a very positive thing. It typically comes after learning from someone else, or more commonly, acquiring and using some of the excellent "how-to" books on the market. When one grows much of his food, learns to repair most things, and build/wire/plumb what's needed, there's a sense that almost anything can be accomplished if approached in a methodical, thoughtful way.

After numerous good and bad experiences of doing various "skill" tasks formerly reserved for high-priced technicians, a new feeling of personal competence emerges. And it is much more than the "Now I know how to wire a building, and won't need electricians ever again, whoopee," stage.

We Are Junior Partners

Farming quickly makes one realize that the natural processes going on all around are far more precise and competent than anything man knows about. Personal humility follows that realization, the knowledge that we are always the junior partner.

That combination becomes a very powerful paradox — confidence that derives from learning living skills, and humility from watching nature in action.

Implications of the Partnership

Then along comes what some would term a more "spiritual," but inescapable conclusion: that man is a part of that nature he's observing, not apart from it. We, too, are like shoots of grass. We are born, we may or may not flourish, and then we die and our material remnants are recycled to nurture further life forms.

Culturally, we in the United States, especially, have come to believe that there is a grandiose creation out there for us to manipulate in various ways for our presumed benefit . . . but we, the human observers, sitting over on the sideline, are not really kith and kin of the rest.

That presumed separation may be at the root of many of our cultural problems. It also may be the impetus behind the love-at-first-sight so many have with seeing chickens clucking around in our pastures, the orderly beds of vivid green growing things, and the chocolate-cake brown of damp, newly-tilled earth. When all five senses are simultaneously engaged a sixth sense seems to make the connection between the earth, growing foodstuffs, and our inner selves.

Farming can be a life more connected to the essentials of earth, sky and the recesses of one's own soul.

Personal Power

The bottom line for the farmer can be a newfound but humble sense of power. Not the supposed power we sometimes feel over people who work for us, our children, and so on, but the power of knowing that we could be plunked down almost anywhere and could fairly soon happily support ourselves and others by growing things — after detailed observation of local natural conditions, befitting the junior partner in the endeavor — and could meet most of our other needs by building and repairing and relying on our inner resources.

This is in contrast to the very narrow range of perceived options — and terror — typically faced by persons in the general workforce when terminated from a job. What a difference this inner self-reliant sense makes! It may be the highest job bonus possible.

Having mentioned the spiritual dimension of farming, it is worth elaborating just a bit further.

The Big Picture

A thoughtful observer working the land can't escape seeing the awesome intricacy and interrelationship of our whole system. Small farmers tend to have hands and feet on/in the soil most of the time. We become very aware that we do not make the plants grow . . . we, as junior partners, just try to create the right circumstances for them to thrive and they in turn help us and our customers thrive.

The most sophisticated sciences know very, very little of the interaction among the millions of living microscopic players and host of elements associated with the growth of one plant. Association with this life process gives one a sense of awe, a childlike wonder, that helps put other aspects of life in perspective.

Some talk about garden divas or fairies or little people or magical places or angels or God's smile in gardens. Whatever the reality, there is, beyond doubt, a mystical bond between people and the good earth.

That bond, largely disregarded and muted in our culture, is rekindled on the small organic farm.

Now It's Your Turn

Many are amazed to discover that we two middle-aged novice farmers are making a decent living on less than two acres of land. Even more amazing is that the model appears to be replicable by almost anyone, almost anywhere in this country and many others. Simply stated, our farm is just an updated version of the market farms/gardens of all recorded history.

Technological advances, careful planning, and marketing directly to the consumer have reestablished this time-honored format as a basis for the new small family farm.

Our hope is to see all urban areas once again surrounded by these farms: many thousands of tiny oases reestablishing the "knowing link" between society and the land that nurtures it. Large-acreage farmland in the path of urban development cannot economically withstand the onslaught of that development. But our very intensive, diverse crop type of farm can economically compete, and simultaneously create a mutually-beneficial relationship with the nearby urban world.

Go for it. Do it. Be a part of it!

Organic hazelnuts await gathering.

Epilogue

As we approached age 60, we each sensed it was time to move to our next life phase. Our 14 years on the farm have been the best of our working lives, but there arose a need to ponder deeper issues, be with family, read and study more, and give of ourselves in a more consistent way. There is something special about reaching 60. It is truly a time to ripened perspective and opportunities, so we've moved on to a new time and place where, among other things, we can be more a part of our five grandchildren's lives, just as our grandparents were there for us in our young lives.

In the eight years since this book was first published, much has changed in our business:

• A number of good new books have been published about small farms.

• Universities and public interest groups are beginning to support organic and small farming systems.

• Organic is the fastest growing segment of agriculture — no longer on the fringe.

• At Island Meadow Farm, we took on an apprentice; had some day volunteers work and learn with us; hit our target of $40,000+ gross sales without continuing any winter upholstery work; worked in the politics of agriculture; and tried new things (raising a few pigs, using rainwater collection, increasing emphasis on retail flower sales, and more).

Also during those years, three of our four parents passed away, leaving each of us as senior family members. But the life cycle continued as we gained two more wonderful grandchildren, and another is expected to join the East Coast branch of the family within a few weeks.

In light of these new realities, people occasionally ask if there are things we would like to change in this book. Well, there are a few — the color of the cover, for one! And the fact that there turns out to be a good market for "spent hens" in the local Asian community. No more need to kill and compost as described on page 1 of this book.

Apart from that, however, we think this book timelessly outlines the essentials of starting a successful small farming business. It has been well stated by whoever said, "Small farming is not a lifestyle. It is a *business,* which, if properly executed, *allows* a certain lifestyle!" We hope we have emphasized the business elements adequately in this book.

There is one more big additional thing to share: the mutually beneficial way in which our farm transitioned from us to the next farming generation, and how this particular technique might work for other farmers, too. It's called a sale with a "life estate," and here's how it works:

The farmer can sell the entire farm, but retain control over some portion (usually the house and a bit of land) for the rest of the natural life of either the farmer or his/her spouse. The value of using the house over all those future years of life expectancy is very substantial — about half of the current appraised value of the entire farm, in our case. Since we kept control of the main house and a few back acres, we considered it fair to sell the entire farm for half price (to a young man who was our first apprentice in 1997). We stayed in one of the two houses on the property and turned over the other to him, plus most of the land and equipment and the operating farm business, without skipping a beat and even without a down payment.

He thus needed no capital to establish his own ongoing farming business, while we get monthly mortgage payments from him as well as a "free" home to live in or rent out, as we see fit. Upon the demise of both of us (the Gregsons), whenever that comes, he gains legal control over the entire farm, although he will continue to make mortgage payments to our estate if the mortgage has not been satisfied. It is a great deal for both parties, but there are some mutually important financial safeguards to work out, so an attorney and CPA should be involved.

This is a win-win situation that can work for other older farm folk who prefer to continue living on the farm and to have the farm stay in business, but who also need income from that land. And, of course, it can also be a great way for a young farmer to become a farm owner.

As for our new adventure, we recently decided to rent out the farmhouse and move across the state to be close to those five grandchildren — and some of our first activities have been installing new raised beds and a small orchard at our new house, and assisting our children as they excitedly install theirs. Life is good!

— Bob & Bonnie Gregson, Spokane, Washington, May 2004

58

Appendix A
Sources of Materials

COVER CROPS & SPECIALIZED FARM MATERIALS

Peaceful Valley Farm Supply
P.O. Box 2209
Grass Valley, CA 95945
Phone (888) 784-1722
Fax (530) 272-4794
www.groworganic.com

SEEDS

Bountiful Gardens
18001 Shafer Ranch Road
Willits, CA 95490
Phone (707) 459-6410
Fax (707) 459-1925
www.bountifulgardens.org

Johnny's Selected Seeds
955 Benton Avenue
Winslow, ME 04901
Phone (877) 564-6697
Fax (207) 861-8363
www.johnnyseeds.com

Pinetree Garden Seeds
P.O. Box 300
New Gloucester, ME 04260
Phone (207) 926-3400
Fax (888) 527-3337
www.superseeds.com

Seeds of Change
P.O. Box 15700
Santa Fe, NM 87592
Phone (888) 762-7333
Fax (505) 438-4571
www.seedsofchange.com

Seeds Trust
P.O. Box 596
Cornville, AZ 86325
Phone (928) 649-3315
Fax (877) 686-7524
www.seedstrust.com

The Cook's Garden
P.O. Box C5030
Warminster, PA 18974
Phone (800) 457-9703
Fax (800) 457-9705
www.cooksgarden.com

Territorial Seed Company
P.O. Box 158
Cottage Grove, OR 97424
Phone (800) 626-0866
Fax (888) 657-3131
www.territorialseed.com

Totally Tomatoes
334 W Stroud Street
Randolph, WI 53956
Phone (800) 345-5977
Fax (888) 477-7333
www.totallytomato.com

Vermont Bean Seed Company
334 West Stroud Street
Randolph, WI 53956
Phone (800) 349-1071
Fax (888) 500-7333
www.vermontbean.com

ORGANIC AMENDMENTS

Concentrates, Inc.
2613 SE 8th Avenue
Portland, OR 97202
Phone (503) 234-7501
Fax (503) 234-7502
www.concentratesnw.com

Gardens Alive
5100 Schenley Place
Lawrenceburg, IN 47025
Phone (513) 354-1482
Fax (513) 354-1484
www.gardensalive.com

T-TAPE & DRIP IRRIGATION

Drip Works
190 Sanhedrin Circle
Willits, CA 95490
Phone (800) 522-3747
Fax (707) 459-9645
www.dripworksusa.com

Appendix B

1995 Island Meadow Farm Crop Selection and Related Data

Crop	Amount Planted	Deliveries to Subscribers	Unit
Dry Beans	150'	1	1 lb. bag
Seasonal Salad Mix	120 starts/week	22	¹/₂ lb. bag
Head Lettuce	120 starts/week	4	head
Kale	60'	4	bunch
Rhubarb	12 hills	5	1¹/₂ lb.
Rosemary	4 large plants	3	bunch
Thyme	12 large plants	1	bunch
Thyme Plant	—	1	3" pot
Hazelnuts	5 acres	13	¹/₂ lb.
Perpetual Spinach	20'	1	bunch
Chard	15'	1	bunch
Baby Dill	36 starts/month	3	bunch
Lemon Balm	2 large plants	1	bunch
Spearmint Plant	—	1	3" pot
Spinach	40'	2	bunch
Assorted Flowers	400'	5	bunch
Apple Mint	10'	2	bunch
Tarragon	15'	1	bunch
Strawberries	100 plants	3	pint
Basil	60'	8	bunch
Mixed Cooking Greens	volunteer plants	2	bag
Collards	10'	2	bunch
Small and Medium Onions	300'	4	1 lb.
Broccoli	300'	3	head or bag of florettes
Summer Squash	72 plants	14	1 lb.
Oregano	10 plants	1	bunch
Peas	400'	1	2 lb.
Globe Artichokes	16 plants	1	4 heads
New Potatoes	100'	1	2 lb.
Potatoes	200'	(sold at stand late in season)	
Lavender	10 plants	1	bunch

Crop	Amount Planted	Deliveries to Subscribers	Unit
Spearmint	10'	1	bunch
Large Sweet Onions	300'	4	1½ lb.
Raspberries	100'	2	pint
Beets with Greens	240'	6	bunch
Small Cabbage	60'	1	head
Cauliflower	150'	1	head
Kohlrabi	60'	2	head
New Zealand Spinach	60'	5	bunch
Pole Beans	200'	8	1 lb.
Sweet Cherries	3 trees	1	pint
Plums	3 trees	2	pint
Elephant Garlic	60'	1	bulb
Assorted Cucumbers	120'	11	4 each
Tomatillos	60'	3	½ lb.
Bulb Fennel	30'	1	bulbs
Assorted Tomatoes	250 plants	10	2½ lb.
Sweet Peppers	30'	3	2 each
Hot Peppers	60'	4	6 each
Mini Eggplants	30'	1	2 each
Garlic	100'	2	3 bulbs
Sweet Corn	400 plants	2	4 ears
Walnuts	2 acres	1	½ lb.
Mini Pumpkins	50'	1	3 each
Pie Pumpkins	200'	1	1 each
Large Pumpkins	100'	1	1 each
Winter Squash	1000'	4	2+ lb.
Mangels and Greens	60'	1	bunch
Assorted Grapes	200'	2	2 lb.
Shallots	120'	1	4 bulbs

This list includes everything we grew, how much total space was devoted to those crops, and how many deliveries of each item were made to the 38 subscribers. The rest we ate, sold at the farmstand, gave away, or sold to grocery stores/restaurants.

Appendix C

The Most Valuable Periodicals

Acres U.S.A.
P.O. Box 91299
Austin, TX 78709
Phone (800) 355-5313
Phone (512) 892-4400
Fax (512) 892-4448
www.acresusa.com

Growing for Market Newsletter
P.O. Box 3747
Lawrence, KS 66046
Phone (800) 307-8949
Phone (785) 748-0605
Fax (785) 748-0609
www.growingformarket.com

Hortideas
750 Black Lick Road
Gravel Switch, KY 40328
http://users.mis.net/~gwill/hi-index.htm

Organic Gardening Magazine
Box 7307
Red Oak, IA 51591
Phone (800) 666-2206
www.organicgardening.com

Small Farmer's Journal
P.O. Box 1627
Sisters, OR 97759
Phone (800) 876-2893
Phone (541) 549-2064
Fax (541) 549-4403
www.smallfarmersjournal.com

Books

The New Organic Grower
by Eliot Coleman

Chelsea Green Publishing
P.O. Box 428
85 N. Main Street, Suite 120
White River Jct., VT 05001
Phone (800) 639-4099
Phone (802) 295-6300
Fax (802) 295-6444
www.chelseagreen.com

The Organic Method Primer
by Bargyla and Gylver Rateaver

Acres U.S.A.
P.O. Box 91299
Austin, TX 78709
Phone (800) 355-5313
Phone (512) 892-4400
Fax (512) 892-4448
www.acresusa.com

Backyard Market Gardening
by Andy Lee

Good Earth Publications
20 GreenWay Place
Buena Vista, VA 24416
Phone (540) 261-8775
Fax (540) 261-8775
www.goodearthpublications.com

Appendix D

Community Supported Agriculture
Resource Organizations

Indian Line Farm

57 Jug End Road
Great Barrington, MA 01230
Phone (413) 528-8301
www.indianlinefarm.com

ATTRA (Appropriate Technology Transfer for Rural Areas)

P.O. Box 3657
Fayetteville, AR 72702
Phone (800) 346-9140
www.attra.org

Bio-Dynamic Farming & Gardening Association

25844 Butler Road
Junction City, OR 97448
Phone (888) 516-7797
Phone (541) 998-0105
Fax (541) 998-0106
www.biodynamics.com

Community Supported Agriculture

USDA National Agricultural Library
Alternative Farming Systems Information-
Center (AFSIC)
10301 Baltimore Avenue, Room 132
Beltsville, MD 20705
Phone (301) 504-6559
Fax (301) 504-6927
www.nal.usda.gov/afsic/pubs/csa/csa.shtml

Notes